British Geography
1918–1945

British Geography 1918–1945

edited by
ROBERT W. STEEL

The right of the
University of Cambridge
to print and sell
all manner of books
was granted by
Henry VIII in 1534.
The University has printed
and published continuously
since 1584.

CAMBRIDGE UNIVERSITY PRESS
Cambridge
London   New York   New Rochelle
Melbourne   Sydney

CAMBRIDGE UNIVERSITY PRESS
Cambridge, New York, Melbourne, Madrid, Cape Town, Singapore, São Paulo

Cambridge University Press
The Edinburgh Building, Cambridge CB2 8RU, UK

Published in the United States of America by Cambridge University Press, New York

www.cambridge.org
Information on this title: www.cambridge.org/9780521247900

First published 1987
This digitally printed version 2008

A catalogue record for this publication is available from the British Library

Library of Congress Cataloguing in Publication data

British geography 1918–1945.
Includes index.
1. Geography – Great Britain. I. Steel,
Robert W. (Robert Walter), 1915–
G99.B75 1987   910′,941   87–6549

ISBN 978-0-521-24790-0 hardback
ISBN 978-0-521-06771-3 paperback

Contents

vi   *Contents*

Preface

This collection of essays began as a direct consequence of the work that I undertook on behalf of the Institute of British Geographers to prepare a history of its first fifty years. It was suggested to me that, while I was delving into the development of the subject in 1933, the year in which the Institute was founded, and the years immediately before then, I might also attempt an assessment of the position of geography in Britain between the wars. The idea appealed to me for I had been taught in Oxford by J. N. L. Baker who had always impressed upon me and my fellow students the importance of an appreciation of the history of geography. I subscribed wholly to the view that he had expressed in a lecture on 'Geography and its history' given to Section E (Geography) of the British Association for the Advancement of Science in 1955 (Baker 1955:198):

> The history of geography is long and honourable. No geographer need apologise for it or be ashamed of it … it is only when the geography of our day is seen against the background of its history that its present position can be appreciated and its future prospects assessed.

I noted, too, that R. J. Johnston in his *Geography and geographers* had observed that 'although this book is about human geography since 1945, the discussions of that period must be preceded by a brief outline of the nature of the discipline in the previous decades' and he suggested that 1945 'did not mark a major divide in the views on geographical philosophy and methodology' (Johnston 1979: 28). And I was also conscious of the many times in the past when there have been references to a 'new geography'.

It seemed right, therefore, to take note of the valuable foundations of geography laid in the past, and not least of the years between the two World Wars of the twentieth century, upon which so much of the discipline, as taught and practised today, is based. It is also important to appreciate how much was done with very slender means and minimal resources of men and women, money and materials.

But clearly it was not possible for an assessment of inter-war geography to be undertaken by one individual, especially as in my case I graduated only during the last few years of that period. I needed the help of collaborators who, while not willing or able to undertake the task single-handed, were prepared to work with others. So the idea of an assessment of inter-war geography was born. It was emphasized that it was not to be a chronological account nor a series of departmental histories, and that, where appropriate, reference should be made to the concepts of those years – relating, for example, to regional geography – and to the controversies that arose from time to time – between, for example, those who subscribed to 'determinism' as opposed to 'possibilism'. During 1980 and 1981 potential contributors were approached and most of them were very happy to join me, especially as it was made clear that there was to be no strait-jacket for their contributions. They were to write on what they knew and on what they had been involved in during such years as they had been geographers between 1918 and 1945. Some had, unfortunately, to decline for health reasons or because of other commitments, and this explains why there is no contribution specifically concerned with either Scotland or Ireland. All who accepted fulfilled their obligations, and how successful they were in their interpretation of their assignment is for readers to judge for themselves.

What I had not anticipated was the effect of death in a group of senior geographers, the majority of whom were seventy or more years of age. As editor I regard it as a special privilege to have been responsible for three essays, all distinctive and in a variety of ways characteristic of these authors right at the end of their distinguished careers, seeing the light of day posthumously. The three colleagues who have died are K. C. Edwards, E. G. Bowen and S. H. Beaver. Their essays in this volume may not be the best of the many that they published during their long and productive lives, but they tell us a great deal about their authors and of the influences that helped, in the years between the wars, to form their careers and to make them the distinguished geographers that they were.

With the editor of the volume being among the youngest of those collaborating, it was necessary for him to adopt a light touch. My colleagues accepted their assignments most willingly and were very ready to listen to my suggestions; but I did not regard it as proper to cajole them in any way or to insist on there being a standard form for each and every essay. So a varied group of geographers has produced what reviewers will no doubt describe as a mixed bag or a motley collection, as they invariably do of collections of essays, perhaps particularly those published in *Festschrift* volumes. In a sense this is a *Festschrift* of the 1918–1945 period. Readers will be able to note how certain branches of the subject – historical geography, for example – developed; and how much work in physical geography was undertaken – by geologists as well as by geographers. They will learn of the ways in which geographers applied themselves to the investigation of a wide range of problems that led to the Land Utilisation Survey of Great Britain and to the early years of the Ministry of Town and Country Planning established during the Second World War. They will also discover how geographers – some of them contributors to this book – trained themselves or were trained in a world that is very different from that of the eighties, and how all of them were restricted in the number of specifically 'geographical' books available to them and in the opportunities provided for participation in fieldwork or post-graduate research.

Inevitably in a volume such as this there is an emphasis on personal experience and involvement in the development of the subject. A better, more integrated and more comprehensive – and perhaps less discursive – book might have been written twenty or fifteen years ago; but in fact no such book was produced by any of the senior geographers then available, and no one attempted an up-dating of Griffith Taylor's *Geography in the Twentieth Century*, published in 1951. It is unlikely that another book similar to the present one will ever appear since a fair proportion of those still active (and also active in the 1920s and 1930s as teachers or students) are already represented in the authorship of this volume. How comparatively few of the pioneers of the inter-war period are still alive is emphasized by one of the discoveries made during the writing of the history of the Institute of British Geographers. When the volume was published in 1984 there were only eighteen of the seventy-three founder members of 1933 still alive (Steel 1984: 145).

Each reader will make his or her assessment of the state and status of geography in the inter-war years in the light of his or her perception of what the different authors have written. It was thought, however, that it would be valuable to have included in the volume a reaction from selected geographers who, while very much younger than the senior authors, had known most, if not all, of them and who would in consequence appreciate the men and women as well as the geography professed by these geographers. They were chosen with considerable care, and in J. Allan Patmore and David R. Stoddart the editor felt that he had two collaborators whose reactions to the subject of this volume would be useful to us and, we hoped, helpful to the readers of these essays as well. One is an Oxford-trained geographer, a specialist in human geography. His university experience since leaving Oxford has been divided between the Departments of Geography in the University of Liverpool and of Hull. The other is a product of the Cambridge Department of Geography with special interest in physical geography as well as in the history of ideas in geography. David Stoddart in contrast to Allan Patmore has spent the whole of his teaching career in one university, Cambridge, though he has a reputation of being one of the most travelled of British geographers, having carried out scientific work in Sierra Leone, Socotra, Aldabra, the USA and many other countries.

These two geographers could have tackled their task in much the same way and produced not dissimilar essays. Happily – and without discussion between themselves or with the editor – they have chosen to deal with their assignments in quite different ways. Dr Stoddart has given us a valuable review of geographers and geomorphology in Britain between the wars, producing an essay that is a valuable piece of work that could stand alone, without reference to the rest of the volume, but is in particular a commentary on J. A. Steers's survey of physical geography in the inter-war period when he was so closely concerned with many of its most significant developments. Professor Patmore's essay, in complete contrast, stems not from one or two of the essays but from all of them, and in a short but evocative essay he directs attention to the rest of the volume while at the same time underlining many of the special difficulties faced, in their younger days, by those who have drawn on their experiences of the inter-war years as the basis of their reminiscent essays.

In offering this book to readers and reviewers, and to all who appreciate, as the authors of these essays do, the position of geography today in schools, universities, polytechnics and education generally, there is therefore a delicate emphasis on the foundations of the subject as it is in the 1980s that go back many years. None of the writers believe – in contrast perhaps to the views of some of our younger colleagues – that geography suddenly emerged in Britain, or arrived from the USA, in the years immediately following the Second World War. It has had a long period of gestation beginning many centuries ago. Geographers do well to remind themselves often of the words of Richard Hakluyt who, in giving some public lectures on geography in Oxford more than 400 years ago, claimed to be 'the first that produced and showed both the old imperfectedly composed, and the new lately reformed maps, globes, spheres and other instruments of this Art'.

I am very grateful to all my collaborators who have cooperated most willingly and understandingly, and have shown great patience. Each of them would wish to thank those who have helped them with their secretarial skills and in other ways. As editor I am grateful for all the secretarial and other assistance given to me over a long period by, among others, Joan Lewis, Margaret Fox, Christine Williams, Betty Murray and Betty Thomson and for the ready help forthcoming from Elspeth Buxton, librarian of the Oxford School of Geography. In the task of editing, I have, as always, been greatly helped by my wife Eileen, a fellow student in the Oxford School of Geography, and my companion in geographical work and travel at home and overseas, particularly in Africa, for nearly half a century.

Swansea   Robert W. Steel

REFERENCES
J. N. L. Baker (1955), 'Geography and its history', *Advancement of Science*, 12, 188–98.
R. J. Johnson (1979), *Geography and Geographers*.
R. W. Steel (1984), *The Institute of British Geographers: the first fifty years*.
G. Taylor (1951), *Geography in the Twentieth Century: a study of growth, fields, techniques, aims and trends*.

1   The beginning and the end

ROBERT W. STEEL

To understand the progress of geography in Britain between 1918 and 1945 it is important to appreciate what was the state of the subject at the beginning and at the end of the period under review. However 'new' geography may sometime seem to us in the twentieth century, certainly as a university discipline, it is a subject with a long and honourable history. Its evolution over the centuries has been studied by many writers, both in general terms and in detail, and it is unnecessary to repeat this story even in summary form. Perhaps it is enough to remind ourselves that geography has long been known and practised in Britain. We know, for example, that as long ago as 1187, Giraldus Cambrensis, a Welsh scholar in Oxford, read aloud his *Topography of Ireland* for three whole days in 1187. Nearly four hundred years later, and again in Oxford, Richard Hakluyt, a Student of Christ Church, gave lectures on geography and subsequently produced the series of volumes called *The Principall navigations, voyages and discoveries of the English nation.* Indeed the view of J. N. L. Baker is that the progress of academic geography in Britain during 'the seventeenth and eighteenth centuries is largely concerned with the University of Oxford which during that period led the way in geographical study and accomplishment, and produced one work of outstanding merit' (Nathanael Carpenter's *Geographie Delineated Forth in Two Bookes, containing the Sphaericall and Topicall Parts thereof,* published in Oxford in 1625). During the later eighteenth century and throughout the nineteenth century geographical ideas played an important role in the work of explorers from various parts of the British Isles – James Cook, for example, with his extensive voyages in the Pacific and elsewhere, and David Livingstone, the great missionary traveller in Africa. Significantly the Royal Geographical Society, in large measure an extension of the African Association established during the latter years of the eighteenth century, was founded in 1830, and one of its first Secretaries, Captain Alexander Maconochie, was given the title of Professor of Geography at University College London between 1833 and 1836 (Ward 1960). Also during the nineteenth century a number of important books on geography were published – perhaps of special influence was Mary Somerville's *Physical Geography*, which first appeared in 1848 (Baker 1948).

The growth of the subject as an academic discipline worthy of study in universities matters most to our understanding of the state of geography between the World Wars of the twentieth century, in which century the most important developments have been concentrated. Much of this progress can be traced back to the publication in 1886 of Scott Keltie's report on geography, commissioned by the Royal Geographical Society (Keltie 1886). Among the consequences of this survey was the interest shown in both Oxford and Cambridge, with much support and encouragement (financial and otherwise) from the Royal Geographical Society. Of special significance was the appointment of H. J. Mackinder to a Readership in Oxford in 1887, with comparable developments in the University of Cambridge. The growth of the subject in schools was also marked by the founding, by a group of schoolmasters, mainly in public schools, of the Geographical Association in 1893.

But although Oxford and Cambridge had the opportunity of leading the field in the development of geography, some events elsewhere were more significant, or at least of equal importance, such as the appointment of L. W. Lyde, a classicist, to a Chair of Economic Geography at University College London in 1903, and the arrival of P. M. Roxby, an Oxford-trained historian, in the University of Liverpool to encourage teaching in the subject. Roxby's presence there led on to the creation of the John Rankin Chair of Geography in 1917 at the same time as the Honour School of Geography in the Faculty of Arts – the first of its kind in any British university – was introduced (Steel 1967: 5). In the following year the University College of Wales, Aberystwyth, introduced honours schools in both Arts and Science, this coinciding with the appointment of H. J. Fleure to the new Gregynog Chair of Geography and Anthropology. Very shortly afterwards honours courses were established in some of the colleges in the University of London where in 1922 Sir Halford Mackinder and J. F. Unstead were elected to Professorships of Geography at the London School of Economics and Birkbeck College respectively. Much was happening, therefore, in the geographical world in the period immediately following the Armistice of 1918, and geography was clearly well poised for development and expansion.

Three of the contributors to this volume have concerned themselves in various ways with the state of the subject from 1918 onwards. T. W. Freeman's *A History of Modern British Geography*, published in 1980, includes two chapters specifically on the inter-war period as well as a chapter on 'Geography in war and peace' (Freeman 1980); H. C. Darby has an essay on 'Academic geography in Britain, 1918–1946' in the special jubilee number of the *Transactions of the Institute of British Geographers*, published in 1983 (Darby 1983). D. R. Stoddart has recently published his *Geography and its History* (Stoddart 1986), and in it he presents an interesting view of the subject from the point of view of someone who was born only a few months before the outbreak of war in 1939. It is superfluous to repeat what these colleagues have written in recent years. It is enough to underline that in 1918 there were more departments than there had ever been before though most of these were young, and such honours schools as there were were as yet untried, although they speedily attracted a fair number of undergraduates. Other universities were also showing an increased interest in the subject though all the geography departments in existence were small with minimal resources in terms of physical space, staff, both academic and ancillary, equipment and resources. And despite the headstart given to Cambridge and Oxford thirty years before, there was still no Cambridge Tripos and no Oxford Honour School. War-time activities of geographers had been important although the number was small (infinitesimally so in comparison with the considerable numbers of trained geographers who were available during the Second World War). Important work was done by the Section of the Naval Intelligence Division of the Admiralty, under the direction of H. N. Dickson (a lecturer in Oxford from 1899 until he became Professor of Geography at University College Reading in 1906), and some fifty handbooks and manuals on different countries together with about 130 short geographical reports were produced. But the nature and scope of this output can hardly be compared with the publication of Geographical Handbooks during the Second World War though then there was, of course, an army of geographers and others to undertake the work compared with the small group responsible for this similar activity in the First World War. This, then, was the situation in which geography embarked upon its course of progress during the inter-war period, and it is against this background that many of the essays in this volume should be seen. Several of the authors have written with first-hand experience of the problems of small departments, broad syllabuses and very meagre resources. That so many of the essays can indicate remarkable progress over the decades that followed the end of the War in 1918 is noteworthy and is in itself a justification for a volume of this nature. It was a modest beginning to a new chapter in the history of British geography.

Just as the post-war period was beginning, Cambridge introduced its Geographical Tripos and in Oxford Professor C. H. Firth produced a forceful pamphlet, 'The Oxford School of Geography', which argued the case for an honours school. These two developments suggested ways forward for the subject though in fact it was only some years later that there were really significant developments. The Cambridge Chair of Geography was not established until 1928, and that in Oxford, coinciding with the introduction of an Honour School, only in 1931. There was an interesting development in the use of geographers in 1918 when A. G. Ogilvie was sent to the Geographical Section of the General Staff, in which capacity he went to Versailles as a member of the British Delegation. Among his activities was a survey of southern Macedonia, and he saw the task of the map makers at the Versailles conference as very much the delineation of boundaries that would be least likely to make trouble later. It is generally felt that the fact that the principles underlining the territorial decisions of the Peace Conference were partly geographical was largely due to the work of the American 'Inquiry' which was centred at the house of the Geographical Society in New York.

Many of the essays in this volume highlight the developments in different places and in different branches of the subject but it is appropriate here to refer to some of the peaks for the subject as a whole that occurred during the twenties and thirties. Thus, in 1928 the International Geographical Union held its conference in Cambridge, and a volume was produced under the editorship of A. G. Ogilvie, *Great Britain: essays in regional geography*. Many years later this volume was described as 'a testimony to the Britain and to the geographers of its period' (Mitchell 1962: xi); and H. C. Darby has commented (1983) that 'the mere fact that the Congress met in Britain at the time sustained and invigorated the growing subject. There were 21 departments of geography in existence when the book was being prepared, and another three were added before the year 1928 was over. It had been a fine summer, and when the geographers returned to their universities they did so with their heads held higher than before' (Darby 1983: 16). In 1930 the Royal Geographical Society celebrated its Centenary, and suggested (despite the economic difficulties of the time) various ways in which it might develop in the coming years and H. R. Mill produced his valuable record of the Society (Mill 1930). Shortly afterwards an Essay Prize was established to encourage undergraduates reading geography in British universities to write with a view to publication in the *Geographical Journal*. The Geographical

Association was increasing its membership quite considerably, and perhaps there was significance in the changing of the title of its journal from *The Geographical Teacher* to *Geography* in 1926. In 1933 a small group of academic geographers came together to start, in a very modest way, the Institute of British Geographers, with but seventy-three founder members but which, a little over half a century later, has a membership exceeding 2,000 (Steel 1984).

All in all the situation in 1939 was vastly different from that of 1918. There were departments of geography in nearly every British university and the number of sixth-form pupils studying the subject had increased enormously. There were more potential recruits for geography in universities, and research prospects were increasing in some universities. A large number of volunteers who had graduated in geography were working for the Land Utilisation Survey of Great Britain under the general direction of L. D. Stamp – undoubtedly one of the major achievements of the subject during the years between the wars (Stamp 1948). Significantly there are references to the work of the Survey in a number of the essays that follow. There were also many geographers whose skills acquired in the university equipped them for service to the nation during the Second World War both in the Services and in the Ministry of Town and Country Planning which was created in 1943 and depended in large measure upon professional geographers for its first members.

If this was the beginning of the inter-war period, what can be said of the position of geography at the end of it? All the contributions reveal in one way or another the growth of certain departments or the enhancement of the subject in particular universities or the progress made in certain branches of the subject as well as in geography as a whole. Some individuals play a very special role in these developments, and it is noteworthy that some names occur repeatedly throughout the volume, often quoted by more than one writer – in alphabetical order J. N. L. Baker, R. O. Buchanan, H. J. Fleure, P. M. Roxby, L. Dudley Stamp, Eva G. R. Taylor and S. W. Wooldridge, to mention but a selection of the pioneers of the subject as we know it and profess it today. Interestingly and significantly each of these geographers graduated in another discipline – this in complete contrast to the writers of these essays, all of whom had geography as the dominant, if not the exclusive, subject in their initial degrees.

Shortly after the end of the Second World War one of these pioneer geographers, Eva Taylor, wrote of the sudden rise in geographical prestige which occurs in war time when 'geographical intelligence of every kind ... becomes vital' (Taylor 1947). Geographers certainly played a prominent part in many fields of national life during the period 1939–45. There was never a comprehensive review of the war-time work done by British geographers (Buchanan 1951; Wilson 1946) though very belatedly the Royal Geographical Society is attempting (in 1985–7) to compile such a record by drawing on the knowledge of those geographers who were involved and who are still alive in the mid-eighties. The fact that geographers could contribute as they did in so many different spheres, both military and non-military, resulted from the training that they had received through the patient and painstaking work done in individual departments and by individual geographers during the years between the wars that are surveyed in this volume. The geographers helped to lay the foundations for future development, providing the springboard, as it were, for the quite remarkable expansion of geography in the universities immediately after the Second World War. Nevertheless, while stressing the importance of what had been accomplished between the wars, one must not under-estimate the achievement of those who worked so hard immediately after the end of hostilities. Perhaps it is not inappropriate to quote here from the history of the Institute of British Geographers:

> The War had destroyed (or at least very seriously disrupted) much of the university system as it existed in Britain in 1939, so that many universities were in considerable disarray, and it would have been easy for them, and their much depleted staffs, to have concentrated wholly on the teaching of the greatly increased number of students, including many returned ex-servicemen, to the total exclusion of research activities and the building up of their disciplines. Departments of geography, never large in the inter-war period, had lost many of their staff. (Steel 1984: 23)

Moreover this remarkable recovery of geography, in both teaching and research, had to be achieved without the help of a number of geographers in academic appointments who had become civil servants in the Ministry of Town and Country Planning. Again, to quote from the history of the IBG: 'in the immediate aftermath of the war, it was by no means clear what was likely to happen to disciplines such as geography, and the chaos in universities as ex-servicemen flooded back to complete their university courses, or, with the help of ex-service grants, to begin their higher education, created many problems and uncertainties in universities throughout the country' (Steel 1984: 24).

Geography did in fact survive the immediate problems of the post-war years. The Institute of British Geographers was re-created, the membership of the Geographical Association increased very markedly and included a fair number of geographers who were not schoolteachers.

The Royal Geographical Society flourished and very considerably extended its activities. A reconstituted National Committee for Geography, with professional geographers in a majority for the first time ever, was established under the aegis of The Royal Society. The number of undergraduates reading geography rose remarkably and university departments increased in number, in size and in quality. It was at this stage that the authors of this volume began to be elected to professorships, and in the following years all the senior authors, apart from the one who was a civil servant, became holders of Chairs of Geography in a variety of universities. It was these departments in the majority of British universities that provided the seed-bed of fertile soil in which new ideas that came from elsewhere – notably the USA and Sweden – were developed, very rapidly in some places, rather more slowly elsewhere and sometimes with considerable scepticism if not positive resistance.

That story of post-war growth and the great expansion of geography during the sixties lies outside the scope of this volume, and is better written by those who are still active in the subject and who have experienced all of this development at first-hand, often as students, then junior staff members and later as holders of key senior posts where they have been very influential, in universities and polytechnics, during the forty years since the end of the Second World War.

At the beginning of the twentieth century H. R. Mill wrote 'we sometimes hear of the New Geography but ... it is more profitable to consider the present position of geography as the outcome of the thought and labours of an unbroken chain of workers, continuously modified by the growth of knowledge, yet old in aim, old even in the expression of the ideas that we are apt to consider most modern' (Mill 1901: 701). Much more recently Peter Gould in his *The Geographer at Work* (1986) has discussed whether geography has passed through an evolution or experienced a revolution in the years since 1945. Whichever view an individual reader adopts, he or she must recognize some continuity – perhaps especially in geography as it has continued through the years and indeed the centuries. In this way these essays, widely different though they may strike the reader in objective, scope and even style, give an indication of geography's strengths and weaknesses in the inter-war years, and it may help them to appreciate both the range and the excitement of a subject that is professed today by, literally, hundreds, even thousands, more than in the 1918–45 period. Happily most of them – even including perhaps those who have not succeeded in finding appropriate appointment for geographers to date since their graduation, and irrespective of the generation to which they belong – are well content to regard themselves

8   *Robert W. Steel*

as involved in what S. W. Wooldridge, as great an expositor of the Scriptures as he was of his geomorphological ideas, referred to, with legitimate adaptation of one of the Pauline letters of the New Testament, as 'the high calling of geographer' (Wooldridge 1950: 11).

8 REFERENCES

8bibliography segment

8
J. N. L. Baker (1935), 'Academic geography in the seventeenth and eighteenth centuries', *Scottish Geographical Magazine*, 51, 129.
  (1948), 'Mary Somerville and geography in England', *Geographical Journal*, 111, 207–22.
R. O. Buchanan (1954), 'The IBG: retrospect and prospect', *Transactions of the Institute of British Geographers*, 20, 1–14.
H. C. Darby (1983), 'Academic geography in Britain: 1918–1946', *Transactions of the Institute of British Geographers*, 8 (N.S.), 14–26.
C. H. Firth (1918), *The Oxford School of Geography*.
T. W. Freeman (1980), *A History of Modern British Geography*.
P. Gould (1986), *The Geographer at Work*.
J. Keltie (1886), Report to the Council of the Royal Geographical Society. *Report of the Proceedings of the Society in reference to the improvement of geographical education*, 1–156.
H. R. Mill (1930), *The Record of the Royal Geographical Society, 1830–1930*.
J. B. Mitchell (ed.) (1962), *Great Britain: Geographical Essays*.
A. G. Ogilvie (ed.) (1928), *Great Britain: Essays in Regional Geography by Twenty-Six Authors*.
L. D. Stamp (1948), *The Land of Britain: Its Use and Misuse*.
R. W. Steel (1967), 'Geography at the University of Liverpool' in R. W. Steel and R. Lawton (eds.) *Liverpool Essays in Geography: A Jubilee Collection*, 1–23.
  (1984), *The Institute of British Geographers: The First Fifty Years*.
D. R. Stoddart (1986), *Geography and its History*.
E. G. R. Taylor (1947), 'Geography in war and peace', *Advancement of Science*, 4, 187–94.
R. G. Ward (1960), 'Captain Alexander Maconochie, R. N., K. H., 1787–1860', *Geographical Journal*, 126, 459–68.
L. S. Wilson (1945), 'Some observations on wartime geography in England', *Geographical Review*, 36, 597–612.
S. W. Wooldridge (1950), 'Reflections on regional geography in teaching and research', *Transactions of the Institute of British Geographers*, 16, 1–11.

2   Geography during the inter-war years

T. W. FREEMAN*

'Far be it from me,' wrote H. J. Fleure in 1916, 'to think of suggesting an Act of Uniformity as regards geographical method ... the adaptability of the subject to the teacher's talents and opportunities is greater than that of most subjects ...' (Fleure 1915–16). He was writing at a time when education in general, and university education in particular, was apparently static through the misery of war but was about to experience a vast expansion, especially in secondary schools and universities. For all this the foundation had been laid in the new grammar schools and in the small departments of geography and other supposedly 'new' subjects. The advance was to come when Honours courses were provided, of which the first was at Liverpool in 1917 in the Arts faculty, followed a year later at Aberystwyth, in geography and anthropology, in both the Arts and Science faculties, and also in London in 1918, and by Cambridge and Leeds in 1919. No directive was given from government or from any national organization of geographers on the content of courses, so it is hardly surprising that they were largely a reflection of the views and tastes of the geographers who had become heads of departments, mostly working with one or two junior colleagues whose work in some cases was supplemented by courses given in other departments such as physical geography by geologists or the history of ancient geography by classical scholars.

No problems arose through association with classical scholars of whom one, J. L. Myres of Oxford, was a very firm supporter of the Geographical

* Thomas Walter Freeman (b. 27 December 1908) graduated in geography (honours) in the University of Leeds in 1930. He was an assistant lecturer in the University of Edinburgh from 1933 to 1935 and was a lecturer at Trinity College, Dublin, from 1936 to 1944 when he was promoted to a Readership. In 1949 he was appointed Reader in Economic Geography in the University of Manchester and was given the personal title of Professor of Geography in 1974. He retired in 1976.

Association and of geography in general. For a time he acted as external examiner in geography in the University of Liverpool where, from 1907 to 1910, he had been Gladstone Professor of Greek and Lecturer in Ancient Geography before becoming Wykeham Professor of Ancient History in the University of Oxford. But the control of the teaching in physical geography by geologists was against the wishes of most (though not all) geographers, including Halford J. Mackinder; though L. W. Lyde at University College London was happy to discard geomorphology as a 'mere morbid futility' to his geological colleagues and to concentrate on regional human geography of his own distinctive vintage. To S. W. Wooldridge and others a sound regional geography was inevitably based on a thorough appreciation of physical geography, on which the work of W. M. Davis and others, including Emmanuel de Martonne in France and Charles A. Cotton in New Zealand, gave enlightenment. The battle between regionalists and systematists was foreshadowed during the inter-war period but many physical geographers regarded their research as a primary contribution to a regional synthesis, though this did not preclude concern with geomorphology for its own interest and value.

Before the time when Honours graduates were available, the demand for suitably trained teachers of geography had been met partly by diploma courses, such as those in Oxford from 1899, with summer schools arranged by various universities or by the Geographical Association. Partly through the tactful support of Mackinder the Royal Geographical Society saw little reason to fear rivalry from the activity of the Geographical Association. Rather they hoped that the new body, founded in 1893, might be more successful than they had been in advancing school education in geography. The Royal Geographical Society remained fully conscious of its concern with exploration, for the mapping of the world, for the mathematical and historical aspects of cartography, and for the study of the British Empire. Generally it was chary of venturing into political geography, but fortunately not always: it welcomed a highly contentious paper on possible post-war European boundaries by L. W. Lyde, who had been professor of geography at University College London from 1903 (Lyde 1915). It also welcomed Marion Newbigin (editor of the *Scottish Geographical Magazine*) when she gave a thoughtful paper on 'Race and nationality' two years later (Newbigin 1917). But after the War the new countries were far more adequately treated in the American *Geographical Review*, vastly improved in content and presentation from 1916 by the vigorous editorship of Isaiah Bowman, and in the *Annales de Géographie*. The old, safe study of exploration, cartography and physical geography prevailed while those who were looking for new fields

of enquiry into the varied aspects of human geography had to look elsewhere for the publication of their articles.

The prophetic voices of geographers

Many students were fired by the enthusiasm of the more charismatic
teachers who came into their golden middle years during the inter-war
period. The early honours courses, notably at Liverpool, Aberystwyth,
Cambridge and London, attracted eager students of whom a number
became university teachers privileged in their turn to establish new
honours courses. In Liverpool the emphasis was on historical and regional
geography for P. M. Roxby (an admirer of A. J. Herbertson, whose teaching was broader than his published work) had made his mark in Oxford
as a student of history, adding to his first-class degree in 1903 the Gladstone Memorial Prize for an essay on Henry Grattan (Freeman 1981).
He was deeply imbued with the scholarship of Vidal de la Blache and
other French geographers and was a particular admirer of the *Tableau
de la géographie de la France* (de la Blache 1903). Of his work on England
the essay on East Anglia in *Great Britain: Essays in Regional Geography*
(Ogilvie 1928) epitomized his view that a physical unit may become an
economic unit, reflecting in its agriculture the inherent qualities of its
soil, climate and other physical attributes. Never attracted to town study,
and indeed always a countryman at heart, Roxby supported the Regional
Survey Association (founded in 1914), which looked forward to the eventual replanning of Britain on a basis of thorough understanding of each
local environment. Roxby naturally approved of any sensible effort to
create a better world, and in teaching on race, then a subject quite generally included in geography courses, he was deeply concerned that all
prejudice should be removed from the student mind. His great love for
China and his various articles on it (none of which appeared in the *Geographical Journal*) were a fine addition to the not-too-abundant literature
of the time in British geography. Roxby and a few others helped to impart
a tradition of good writing in British geography. In research, however,
he accepted some sources as primary that others might regard as secondary, notably the *Christian Occupation of China* of 1922 which was the
basis of much of his work on population. Some of us turned eagerly
to any new article he wrote (few in number) as the clarity of exposition
and the elegance of expression were in such marked contrast to the tur-
gidity of many texts of the day (not including, however, Bowman's *New
World*, 1921, 1924, 1928).
Roxby made his case with power, as a fine preacher or determined

advocate might do. Fleure, both in speech and in writing, had an entirely different technique: he was a persuader of considerable charm. Gently, almost confidentially, he led the listener or the reader on, suggesting, unfolding a hypothesis in which the truth might be found, drawing out the relationships between people and their environment, using but modifying many of the findings of physical anthropology current at the time on race, finding in racial qualities – Nordic, Alpine, Mediterranean – some explanation of personality and human achievement, strength and infirmity. To Fleure a broad view was natural, for he had come to the study of mankind and environment from geology, botany and zoology, with a full appreciation of the evolutionary sequence of life (though without favouring the social Darwinist outlook of Herbert Spencer). Like many of his contemporaries, he regarded the aspirations of Patrick Geddes with not uncritical respect. In a recent study W. Iain Stevenson says that Fleure 'found the inspiration of much of his geographical thought in the sparkling mind of Patrick Geddes' (Stevenson 1978), while in his turn Geddes 'found in Fleure a mind that equalled his own in breadth of interests and synthetic power ... Originally a zoologist, Fleure shared Geddes's background in the natural sciences, but ... he too became dissatisfied with the abstract and mechanistic outlook of science.' In anthropology and geography, 'the discipline of actuality', he saw 'opportunities for the development of a more satisfying humane viewpoint'.

A crucial question of the time was the place of geography in the academic spectrum of knowledge. Leaving aside the disdainful attitudes sometimes held by academics for specialists in other fields (anyone who has lived in a university will need no enlightenment on that), some geographers in all sincerity questioned whether the broad aspirations of Fleure were beyond human achievement. Was progress made through generalization, inductive rather than deductive argument, intuition, hypothesis based on a slender factual foundation, or was it better to proceed by detailed observation and research? Just as the rigid scientist wanted to find the truth by experiment, so the historian searched every possible documentary source. Field evidence, used plentifully by natural scientists but less enthusiastically by historians (with significant exceptions), needed critical interpretations; but the broad sweeps of Fleure and others, though undoubtedly stimulating and likely to lead to a new and amended synthesis, appeared to some people to be inadequately grounded in detailed research. What enjoyment, for example, was given to the later workers who demolished the theory of 'valleyward movement of population' in prehistoric times, originally put forward by Fleure and W. E. Whitehouse in 1916, such as S. W. Wooldridge and D. L. Linton who showed that the early settlers were more interested in cultivable and well-drained soils, such as loams, than in altitude (Wooldridge and Linton 1933). Some ideas of the time were stimulating, such as those of Fleure's 'human regions' (Fleure 1919) which, though hard to define and virtually impossible to map, had the merit of carrying the student mentally far beyond the limited but useful climatic regions of Herbertson. Fleure in his retirement said that in his day geographers were in an experimental phase, looking forward to new possibilities of geographical work in which new methods of research and new interpretations of data would arise.

Roxby and Fleure were men of broad if differing vision, sharing a reluctance to define geography in rigid terms or to assign to it precise and clear limits. A. G. Ogilvie in Edinburgh was a man cast in a more cautious Scottish mould, imbued with a deep sense of the relations between physical and human geography in his regional work (splendidly seen in his essay on Central Scotland in the volume *Great Britain: Essays in Regional Geography*, which he edited in 1928) but eager to encourage systematic geography as the essential basis for regional synthesis. He was a particular enthusiast for detailed map work in regional study. He had acquired a considerable knowledge of French and German research before the First World War: during the war his military service gave him a fine appreciation of the geography of the Balkans; and in the immediate post-war years he had a period of work at the American Geographical Society, which was then beginning its mapping and geographical survey of the South American continent. No British geographer of his time had so wide an experience of American and continental European trends in the subject although in Edinburgh he threw all his energies into the encouragement of research on Scotland with the educational advance of the subject in schools.

Geography's problems of definition were a gift to enemies of the subject in the universities, including some historians, economists and geologists to whom the idea of 'breadth' meant superficiality and the then current idea of a 'bridge subject', drawing material from the natural sciences and the humanities with apparent impunity, seemed to make geography a 'robber economy', with no core of its own except an eagerness to comment on the 'geographical aspects' of almost anything that could be described as a natural or a human distribution. In fact geographers were facing a fundamental problem that was to remain a challenge to their work. They saw the challenge of studying 'man and environment'; they appreciated that everything happened in space as well as in time, 'somewhere and somewhen'; but the constraints as well as the opportunities of environmental circumstances were complex and clearly what had been achieved in differing human environments (or regions if that term were used) was an expression of human activity from the first beginnings of settlement. In China Roxby was dealing with a land of intensively settled lowlands with small farms depending on irrigation, and with a drainage system so moulded and modified by human effort for hundreds, even, in some areas, thousands, of years that it was impossible to discern the initial system of natural drainage. Conversely, in much of the world only the most elementary imprint had been made on the natural landscape, even in China, for as Roxby and others extended their work they found that the Chinese were still carrying their agricultural methods forward into previously unoccupied areas, creating a new farmed landscape where a jungle had existed before.

The vision of the world

Exploration, colonial survey, the settlement of new lands, especially the United States of America and Canada, Australia and New Zealand, had opened up the world and given to surveyors and map-makers a challenge that they were eager to accept. Their topographical maps and increasingly informative atlases were a valued contribution to the knowledge of the whole world, and the accumulation of climatic data made possible a climatic regionalization, especially by German workers and in Britain notably by A. J. Herbertson (Herbertson 1905). Later generations could enjoy denigrating such schemes of world division as Herbertson's 'natural regions' as 'just climatic' but the relation of vegetation to climate and soil, and the revelation that tundra could exist on mountains at the equator, were fascinating discoveries before they became commonplace facts of observation. Emphasis on 'distributions' was general but with it there was also an emphasis on origin and developments, and on environmental change from one epoch to another, in some areas made steadily more explicable through research on glaciation and on the history of climate.

Regrettably Herbertson regarded climate as unchanging but in America Ellsworth Huntington was arguing that changes had occurred, even if his evidence was not always convincing. But the deeper problem was the relation of human activity to such changes, on which some enlightenment came when climatic statistics showed the variability of rainfall, especially in areas marginal for settled agriculture. Along with the study of natural environment went the study of the world's population. Words and phrases that later provided an easy target for scornful wisecrackers were prevalent, such as 'human response' to 'environment', meaning the physical environment; but every farmer knew that some crops were best suited to certain fields while on a wider countrywide or even continental scale the pattern of land use bore a relation to inherent physical conditions, however great the modifications and developments made possible by such scientific advances as the emergence of quick-ripening wheats or soil transformation by manuring and chemical fertilization. That two or more blades of grass might be made to grow where one grew before, or that rich arable crops could be harvested on downland previously thought to be suited only to sheep pasture, seemed as great a human triumph as the use of irrigation to make the desert blossom as the rose. There was, it seemed, no limit to the human capacity to make and remake the earth. It was easy to forget that there were areas of the world ruined by deforestation or devastated by soil erosion on marginal land from which the exposed topsoil had been swept away by desiccating winds. Not always was the penetration of the wilderness rewarded with success; on this, later, Isaiah Bowman was to comment that 'man takes the best and lets the rest go'.

An approach which seemed natural during the inter-war period was to consider the agricultural life and to work forward to villages and country towns as the service centres for the countryside. The emphasis was on area and land use within the countryside: in Britain L. Dudley Stamp's plans for a land-use survey from 1929 fell on receptive ears because he asked people to do what seemed to be logical and sensible (Stamp 1947). Despite the clear indication that towns were spreading and growing into one another as conurbations, the threat to the countryside appeared to be an aesthetic rather than a financial problem since farmers were only too eager to sell off building sites at a time when prices for crops were low. Bungalows were a more remunerative crop than beans and it was assumed that Britain would be able to import all the food that was needed, particularly from the Commonwealth. Government paid little attention to the Land Utilisation Survey until 1939, when with the outbreak of war it asked its officials for all possible information as the country was faced with the need to increase domestic food production.

In the early 1930s the emphasis in teaching was on landscape. In most cases this was not derived in any direct way from German geographers but rather from admiration of the French school, for in the 1930s the total written contribution of British geographers was small, American geographers had not reached the eminence they were to achieve later, and most British academics were able to read French easily. The volumes of the *Géographie Universelle* appeared from 1927, and Vidal de la Blache's *Tableau de la géographie de la France* (1903) with the *Principes de géographie humaine* (1922; translated into English, 1926) were widely read and praised, especially by Fleure and Roxby. Other French works, such as those of Jean Brunhes, were also well known and the *Annales de géographie* was in the front rank of world geographical journals. France seemed to be a fortunate land, having at the time approximately half its population on the land and half in the towns, and far less beset with unemployment than either Britain or Germany. But in fact France was facing a population problem with its dependence on immigrant labour regarded by some observers as excessive and to which the Germans of the Nazi period paid particular attention. There were other ominous political signs for those prepared to see them: in the Far East Japan had invaded Manchuria, and one wondered how long China could remain free from military conquest and what in time was to happen to the stable way of life that Roxby so lovingly and eloquently described ('a civilization rather than a nation').

Although Fleure and Roxby had the ability to charm and even captivate their listeners and readers, there were other siren voices among geographers, particularly C. B. Fawcett, whose gentle presentation of his work was seen in papers on population, especially in his discussion of British conurbations in 1931 (Fawcett 1932). Like some post-1945 geographers he saw the need for a study of demography on a local and regional basis and his colleagues, including R. E. Dickinson and A. E. Smailes, followed various urban studies which at the time were novel and welcome. Dickinson was constantly urging that more attention should be paid to the German contribution on urban and regional geography, while Fawcett wrote with circumspection on political geography including studies of international frontiers (1918) and local administrative boundaries (1919), as well as of the British Commonwealth, which he regarded as one of the greatest influences for peace in world history (1933). Fawcett took a more restricted view of what might be achieved than Fleure or Roxby both of whom to him, as to many other people, appeared to be taking so broad a view that it was almost beyond human comprehension. His attitude to history was that what had happened in the past hundred years was of more significance than all that had occurred before. On one point, however, all geographers were in general agreement for, whatever their approach, they wished all students to have some kind of a world view. Fawcett's first year teaching included an engaging presentation of the world distribution of climate, following Herbertson, along with agriculture and population; Fleure's approach was more anthropological and social; while Roxby's was broadly human under the inspiration of Vidal de la Blache. Deeper differences lay in the fact that Fawcett did not regard the teaching of anthropology, or even 'racial geography', as belonging to geography at all.

While all were attracted to the idea of 'regions', Roxby was a devoted supporter of the French 'pays', and Fleure remained attached to his 'human regions' discussed in his paper of 1919 which appeared in the *Scottish Geographical Magazine*. This paper had been rejected by the *Geographical Journal* which in fact never published any of Fleure's work. Fawcett, by nature more cautious than his colleagues, approached regional geography in practical terms as possible administrative districts to replace outmoded counties delineated hundreds of years before. His work was to receive careful attention after 1945 when the long discussion of the revision of county boundaries began which culminated in the compromise arrangements of 1974. Despite all the lip service paid to regional geography, little was made in Britain of the work of J. F. Unstead which, indeed, was much better known in Germany (Unstead 1933). There was, however, the successful textbook of Unstead and E. G. R. Taylor, first published in London in 1910, in which the idea of the region is given as 'those outstanding differences of relief, climate and natural resources which have had the most marked influence upon the development and activities of man' (Unstead, 1927 edition, 237). Unstead and Taylor with their use of natural resources show an advance beyond Herbertson's basically climatic scheme.

Veneration for regional geography (which reached its peak in Richard Hartshorne's *Nature of Geography* (Hartshorne 1939)) was such that a number of geographers valued their geomorphological researches as a contribution to regional and historical geography; an example is S. W. Wooldridge's essay on 'The Anglo-Saxon settlement' in H. C. Darby's *Historical Geography of England* (1936). This was not, however, a universal attitude, for many remained convinced that systematic study, of geomorphology, climatology and human (including political, social and economic) geography, was of value. But there was the inherent view that 'natural' regions were permanent while political boundaries were ephemeral. The most effective but perhaps least regarded riposte to this came in the 1939 address of A. Stevens to the British Association in which he argued that the regional entities of western Europe were man-made: not even a coalfield had a unity if it was divided between two or more powers (Stevens 1939). In all human aspects of regionalization, the situation appeared to be chaotic.

Most British geographers were unwilling to restrict their study to the material features of the landscape, for the appeal to consider people, not only in communities but even as individuals, has never fallen on deaf ears. Few were content to deal only with the material landscape. However, in 1933 P. W. Bryan's *Man's Adaptation of Nature: studies of the cultural landscape* received severe criticism, indeed condemnation, in a review by Eva G. R. Taylor. Her argument was that such restriction was academically unsound for a far deeper approach was needed to give any adequate understanding of existing landscapes, including, for example, the study of origins and development with − for the past hundred years or more in some countries − cartographical evidence and a willingness to consider the effects of modern economic policies, both of governments and of speculative builders or regional planners (*Geographical Journal*, 1933: 81, 352–5). As she said in another review a year later, between the physical and the actual landscape there is always the 'idea', the 'cultural pattern' (*Geographical Journal* 1934: 84, 537).

Shortly before the war P. R. Crowe and R. E. Dickinson wrote on the regional approach in geography (Crowe 1938; Dickinson 1939). To Crowe (and no doubt to many others) it seemed that there was 'probably less community of aim and aspiration among British geographers than among any equivalent group abroad'. He was critical of many of the attitudes and practices of the time, such as the dependence on maps as evidence regardless of other sources, for 'We can read *from* a map only what the cartographer puts into it ... a good map always asks more questions than it is able to answer'. Crowe's view was perhaps crystallized in his statement that 'only by a dynamical study of man's geographical reactions shall we approach the truth'. By contrast, Dickinson argued that 'the geographical study of area has a clear objective in landscape and society, in their association and variations *in area* interpreted in both their genetic development and dynamic relationships ... a definite field of material gives scope, distinctiveness and direction to geographical investigation'. Dickinson finally stated that 'the principal objective, and that which gives unity to the whole, is an appreciation of the changing character of the area directed towards an understanding of its present physiognomy, function and individuality'. There were differences of view between the two writers, especially on the value and meaning of the then current German theories of *Landschaft*, but also some common ground; and at this distance the two papers are interesting as an expression of an aspiration beyond the somewhat simplistic talking points of the time, including 'man and environment', 'material and cultural landscapes', and even 'determinism and possibilism'.

Although on regions geographers spoke with many voices, books on regional geography were in considerable demand for school and university courses. L. W. Lyde, for example, produced a work on *Peninsular Europe* in 1931, followed in 1933 by a vast work on Asia so difficult to read that the present writer still remembers the grim experience of reviewing it. Then there was *North America* by Ll. Rodwell Jones and P. W. Bryan, first published in 1924; some of its chapters were interesting but many were not. L. D. Stamp's work on *Asia* (1929) and Walter Fitzgerald's on *Africa* (1934) were pioneer studies while for the homeland Stamp and Beaver's *British Isles* (1933), strongly economic in approach, supplemented the presentation given in A. G. Ogilvie's *Great Britian; Essays in Regional Geography* (1928). All these books had their limitations, but it was unfortunate that people were so ready to adopt a denigratory attitude, to refuse to see the good points of books written to meet a clear need, even in some cases to talk as if Mackinder's *Britain and the British Seas* (1902) was a classic not only of its own time but of all succeeding time. Mackinder's book in fact received from some people such extravagant praise that an emotional element appeared to be involved. Though some excellent texts came from America, many of their geographers appeared to be obsessed with methodology and the old battles between possibilists and determinists still appeared to be raging.

If, then, many of the books published during the inter-war years proved to be disappointing, there was always the hope that wisdom could be found in papers published in journals. There the biggest disappointment was that few journals published new material, for only rarely was any paper on human (including social, economic and political) geography published in the *Geographical Journal*, the pages of which were devoted mainly to exploration, cartography and, occasionally, geomorphology. The Royal Geographical Society formed a commitee in 1929 to consider publishing such papers (Freeman 1980a: 34–9); by the later 1930s a few of these papers had appeared but many working geographers turned automatically to *Geography*, the *Scottish Geographical Magazine, Economic Geography* and the *Geographical Review*, the most lavishly produced and enterprising of them all, as well as to French and German journals. Later, in 1938, physical geographers launched the *Journal of Geomorphology*, which regrettably survived only to 1942. A major reason for the founding of the Institute of British Geographers in 1933 was dissatisfaction with the work of the Royal Geographical Society but its initial policy was to publish monographs rather than papers (Steel 1984: 56). Only six monographs had appeared before the Institute went into virtual hibernation during the Second World War and only a few after the war, when it was decided to concentrate on the publication of papers. Among the pre-war monographs were Alice Garnett's 'Insolation and relief' (1937) and 'Structure, surface and drainage in south-east England' (1939) by S. W. Wooldridge and D. L. Linton. From the limited response to the opportunity of publishing monographs by the Institute of British Geography and (from 1947) by the Royal Geographical Society, it would seem that few geographers in Britain write this kind of study, of about 30,000 to 50,000 words; though it may be that the higher-degree theses (which may well be the forerunners of monographs) are so clogged with information and references that re-working into readable prose proves to be very difficult or almost impossible.

Crisp and incisive presentation of material was one particular aim of L. Dudley Stamp, who in various articles showed the advance of the Land Utilisation Survey from 1931 onwards. The first twelve of the ninety-two reports appeared between 1936 and 1939, and the rest were in print by 1946. Stamp's *The Land of Britain: its use and misuse* (1947) showed the geographical basis of planning in Britain, made necessary not only by devastation but also by the inevitable pause in erecting houses and public buildings in war-time. What began as an academic exercise, in which anyone able to read a map could take part, became a survey on which a new Britain could be based, part of a forward-looking ethos retained by the British through the darkest war periods. Nothing showed better what geography could do for the community; and it gave a fine stimulus to many geographers who later, either as planners or as academics, were concerned with applied geography. The work of Stamp and his associates stimulated the growth of applied geography, pragmatic in its direct concern with the landscape and visonary in its hope of social and economic regeneration.

Fortunately historical geography also showed a marked advance. The publication in 1936 of *An Historical Geography of England before 1800*, edited by H. C. Darby, showed the possibility of evoking landscapes at various periods. Although later many people, including Darby himself, suggested that an evolutionary approach might also be of value, the 1936 book gave a new precision to historical geography by asking what Britain was like at each chosen time. Naturally some of the fourteen chapters by eleven authors were more convincing than others but in general the work was in accord with academic tendencies in archaeology, for with close field-work, pollen analysis of past vegetation from peat deposits, and other evidence, of which some was geomorphological and geological, it was possible to show not only where, but even how, prehistoric communities lived. Many young geographers of the time were fascinated by the *Corridors of Time* books of H. J. E. Peake and H. J. Fleure which appeared between 1927 and 1936. For later periods documentary sources and maps might be available and, following the pioneer work of O. G. S. Crawford, aerial photography came to reveal many former fields, farms, villages and roads long since disused and forgotten. The modern significance of the aerial photograph, both for archaeology and later history, needs no emphasis.

Historical geography acquired new strength and purpose. H. C. Darby, with others, was working on the Domesday Survey, ultimately to be studied in five volumes published from 1957 to 1967 with a gazetteer in 1975 and a summary volume in 1977. E. G. R. Taylor also drew attention to the concepts of geography held in the sixteenth and seventeenth centuries in *Tudor Geography, 1485–1583* (1930) and in *Late Tudor and Early Stuart Geography, 1583–1650* (1934). There were, too, occasional articles by J. N. L. Baker on the history of geography, such as those on the geography of Daniel Defoe and on academic geography in the seventeenth and eighteenth centuries (1931; 1935), his book *A History of Geographical Discovery and Exploration* (1931), and his collected papers, *The History of Geography* (1963). In 1933 *The Making of Geography* by R. E. Dickinson and O. J. R. Howarth appeared: both these authors were to make substantial contributions to the historiography of geography later. Behind all this enterprise lay the classic work of C. R. Beazley (*The Dawn of Modern Geography*, 1897–1906 3 vols) and E. H. Bunbury (*A History of Ancient Geography*, 1879 and 1883 2 vols), and in several universities courses were given on the history of geographical discovery and exploration. Both Taylor and Baker gave fine service to the Hakluyt Society, which has attracted historians, geographers, explorers, librarians and many others to its membership through the years since its foundation in 1846.

Geography and public policy

Many geographers working during the 1930s hoped for a time when they would be able to comment on issues of public policy, both in Britain and within the Commonwealth. Such papers as Clement Gillman's 'Population map of Tanganyika territory' (Gillman 1936; Hoyle 1981) were indicative of what might be done but few such papers appeared: significantly, after the Second World War, the appeal by E. W. Gilbert and R. W. Steel (Gilbert and Steel 1945) for work on the social and economic geography of colonial territories received widespread support, partly because it was in accord with what many people had thought for a long time. The contributions made by individuals to geographical studies in overseas countries where they settled lies beyond the compass of this paper: others used periods of study leave for such work. Naturally in time the task was handed on to geographers trained in the universities of Commonwealth countries, some of whom were privileged to assist in founding new journals (or perhaps reviving old ones) in which papers based on local research could be published.

Some geographers viewed with impatience the zeal for remote lands although not everyone agreed with the acerbic comment of S. W. Wooldridge that 'the eyes of the fool are on the ends of the earth'. There was certainly in Britain a need for the geography of the here and now, for in the 1930s trade depression had brought widespread misery to coalfields such as South Wales, West Cumberland, and Northumberland and Durham. In the early 1930s the most useful reports available on such areas were written by economists but by the end of the decade geographers eagerly seized the opportunity of collating material on the distribution of the industrial population for the British Government. For England and Wales this work was done by Eva G. R. Taylor and others in association with the Royal Geographical Society (for Scotland it was under the care of D. L. Linton) and the results were presented as evidence to the Barlow *Commission on the Distribution of the Industrial Population* (Freeman 1980b: 139). Naturally there was concern at the time with the decaying industrial 'special' areas, with the vulnerability of industry in the south-east (especially in the Greater London area) to air attack, with the renewal of road and rail communication, with the preservation of the countryside not merely as an amenity but also for its agriculture and forestry, and with the essential replanning of the country. One feature of the enquiry was a strong appeal for a National Atlas, comparable with those of other countries. Taylor became a strong and sensible advocate of national planning during and after the war while retaining and developing her own distinct research interests on the history of navigation and her work for the Hakluyt Society. Perhaps it was fortunate that the Royal Geographical Society (having offered the work of preparing evidence for the Barlow Commission to L. D. Stamp, who had to refuse because of other commitments) engaged Eva G. R. Taylor's services, for in their different but complementary ways she and Stamp became powerful advocates for applied geography both during and after the war. Both were practical people, realizing that applied geography must rest on pure geography. Ideas held by Patrick Geddes and others forty years earlier became an inspiration for action, practical as well as visionary. Many local studies that seemed to be just interesting academic exercises acquired a new significance and at no time was it clearer that 'bread cast on the waters may return after many days'. In the Second World War geographers found many openings for their services, and in the new world of tenuous peace they were contributors to the replanning of Britain. But that is a story to be told by others.

REFERENCES

J. N. L. Baker (1931), 'Geography of Daniel Defoe', *Scottish Geographical Magazine*, 47, 257–69.

(1935), 'Academic geography in the seventeenth and eighteenth centuries', *Scottish Geographical Magazine*, 51, 129–43.

P. R. Crowe (1938), 'On progress in geography', *Scottish Geographical Magazine*, 54, 1–19.

R. E. Dickinson (1939), 'Landscape and society', *Scottish Geographical Magazine*, 55, 1–14.

C. B. Fawcett (1918), *Frontiers: A Study in Political Geography*.

(1919), *Provinces of England: A Study of some Geographical Aspects of Devolution* (also 1960, revised and edited by W. G. East and S. W. Wooldridge).

(1932), 'Distribution of the urban population in Great Britain', *Geographical Journal*, 79, 100–16.

(1933), *A Political Geography of the British Empire*.

H. J. Fleure (1915–16), Comment in *Geographical Teacher*, 8, 89.

(1919), 'Human regions', *Scottish Geographical Magazine*, 35, 94–105.

(1927), See H. J. E. Peake.

T. W. Freeman (1980a), In E. H. Brown (ed.), *Geography Yesterday and Tomorrow*, Royal Geographical Society, 34–7.

(1980b), *A History of Modern British Geography*.

(1981), 'Percy Maude Roxby 1880–1947', *Geographers: Biobibliographical Studies*, 5, 109–16.

E. W. Gilbert and R. W. Steel (1945), 'Social geography and its place in colonial studies', *Geographical Journal*, 106, 118–31.

C. Gillmann (1936), 'Population map of Tanganyika territory', *Geographical Review*, 26, 353–75.

R. Hartshorne (1939), 'The nature of geography', *Annals of the Association of American Geographers*, 29, 171–658.

A. J. Herbertson (1905), 'The major natural regions', *Geographical Journal*, 25, 300–12.

B. S. Hoyle (1981), 'Clement Gillmann 1882–1946', *Geographers: Biobibliographical Studies*, 5, 109–16.

L. W. Lyde (1915), 'Types of political frontiers in Europe', *Geographical Journal*, 45, 126–45.

M. I. Newbigin (1917), 'Race and nationality', *Geographical Journal*, 50, 313–28.

A. G. Ogilvie (ed.) (1928), *Great Britain: Essays in Regional Geography*.

H. J. E. Peake and H. J. Fleure (1927–1956), *Corridors of Time*, 9 vols from vol. 1, *Apes and Men* (1927) to vol. 9, *The Law and the Prophets* (1936);

also vol. 10, *Times and Places* (1956).

L. D. Stamp (1947), *The Land of Britain: Its Use and Misuse.*

R. W. Steel (1984), *The Institute of British Geographers: the first fifty years.*

A. Stevens (1939), 'The natural geographical region', *Scottish Geographical Magazine,* 55, 305–17.

W. I. Stevenson (1978), 'Patrick Geddes, 1854–1932', *Geographers: Biobibliographical Studies,* 2, 53–61.

J. F. Unstead (1933), 'A system of regional geography', *Geography,* 18, 175–87.

J. F. Unstead and E. G. R. Taylor (1910 and later editions), *General and Regional Geography for Students.*

S. W. Wooldridge and D. L. Linton (1933), 'The loam-terrains of southeast England and their relation to early history', *Antiquity,* 7, 179–310, 473–5.

3 Geography in the University of Wales, 1918–1948

E. G. BOWEN*

Geography was taught at the University College of Wales, Aberystwyth, from the days when the College first opened its doors in 1872 – that is, forty-six years before a University Department providing first degree courses with Honours in the subject was established in 1918. All that was given were lectures similar in content to 'the use of the globe', a recital of the names of the chief mountains, the capes and bays and principal rivers, followed by the largest towns of a selected country, and ending with the imports and exports of the country concerned. This represented little more than the content of the geography syllabus found in any primary school in Britain in Victorian times. The lecturers who dealt with the subject had little or no geographical background and took on the task of instruction as a mere 'odd job' imposed upon them by the Principal. Many of them, however, were well qualified in their own special fields to give lectures on geography, as, for example, the first member of staff, Reverend W. Hoskins Abrall. He was a Fellow of Lincoln College, Oxford, and became the first Professor of Classics at Aberystwyth, retaining at the same time his post as vicar of a Herefordshire parish. There were some, however, who gave lectures in geography and geology in the early days who had a wider background as, for example, Leonard Lyell, nephew of the distinguished geologist, Sir Charles Lyell. Lyell did not remain in Aberystwyth long enough to influence the development of the subject generally but he may have initiated the close association between geography and geology that was to remain long after the Department of Geography was established in 1918.

* Emrys George Bowen (b. 28 December 1900) graduated in geography (honours) at the University College of Wales, Aberystwyth, in 1923. He was appointed to a lectureship in Aberystwyth in 1929 and became a Senior Lecturer in 1939. He was elected to the Gregynog Chair of Geography and Anthropology in 1946 and retired in 1968. He died at Aberystwyth on 8 November 1983.

The students were, on the whole, young and immature, finding academic progress slow and difficult. Many sat for the London Matriculation only, and left to become teachers in elementary schools; but the more able remained to take the University of London Degree Examinations externally, before the University of Wales received its own degree-granting charter in 1893. The student intake was very definitely streamlined when Principal T. F. Roberts, almost immediately after his appointment, exerted pressure on the Board of Education (as it then was) to permit the establishment in the University College of Wales of a Degree Training Scheme for prospective teachers in elementary schools. The Board had already, in 1890, permitted such a scheme in colleges of university rank, and Aberystwyth received its go-ahead in 1892. The scheme increased considerably the number of men and women students attending lectures in geography for the Certificate of Education. It was made available to prospective secondary school teachers in 1905. Thereby geography not only became more widely taught in College, but it also did so in the context of the preparation of candidates for the teaching profession. As time passed this was to become very important for both the growth and the content of the geography syllabus subsequently taught at Aberystwyth.

It was to student audiences gathered in this way that a young assistant lecturer, recently returned from a two-year period of research work in Zurich, gave his first lectures in human geography. He was Herbert John Fleure. He had come to Aberystwyth from Guernsey as a freshman in 1897 and graduated with First-Class Honours in Zoology in 1901. He spent from 1902 to 1904 in Zurich engaged on research work in Zoology and acquired a special training in physical anthropology. His wish to study physical anthropology was closely linked with a desire he had acquired in his school days to attempt to understand more deeply the evolution of life on this planet as conceived by Charles Darwin and his followers in the second half of the nineteenth century. Hereafter he always stressed his well-known dictum that 'Man is a part of Nature'. He realized at the same time that the distribution of natural phenomena on the Earth's surface – the mountains, the seas, the variations in temperature and rainfall, the soils and the natural vegetation – were all a part of Man's natural environment, changing somewhat over the ages, and that these were the matters that physical geographers studied. It is not always appreciated that Fleure, in addition, took the fullest advantage of his stay in Europe at this period to deepen his knowledge of the then emerging subject of geography which in Germany and France was ahead, at that time, of contemporary recognition in Britain. When these elements – Man and his Environment – were studied together from an evolutionary point of view, Fleure maintained that here was the essential basis of human geography.

It was with this simple model of human geography in mind that Fleure returned to Aberystwyth in the autumn of 1904, feeling himself fully competent to present it to young teachers in training. In the same year he was appointed to what he has described as an 'odd-job post', namely an assistant lectureship in zoology, botany and geology. When this appointment was offered to him he accepted on the understanding that he would be permitted to pursue work in both geography and anthropology. Herein lie the first steps in the preparatory phase of development of the teaching of geography at university level at Aberystwyth.

The next phase involved Fleure in undertaking a large-scale field investigation into the physical characters of the Welsh people. This was original work of the first importance that was finally published in the *Journal of the Royal Anthropological Institute* in 1916. During the same phase Fleure's efforts as a teacher of geography were strongly brought to the fore by a new development: in 1906 Dr R. D. Roberts, a distinguished pioneer of adult education in Britain, who was concurrently External Registrar of the University of London, persuaded the Royal Geographical Society to make a small grant to enable the University College of Wales at Aberystwyth to establish a lectureship in geography. Dr Roberts was a native of the town and had studied geography at Cambridge, becoming particularly interested in physical geography. He had conducted an extramural course in physical geography in Aberystwyth during the previous year. He was also a member of the Council of the University College and fully aware of Fleure's potentialities.

As was to be expected, the new lectureship in geography had to be established in the Department of Education concerned with the training of teachers, and A. W. Andrews, himself a practising teacher, was appointed to the post on a temporary basis. In 1907 the lectureship fell vacant and Fleure begged for a chance to concentrate on what he considered would be a definite and congenial line of work. The vacant lectureship provided an ideal opportunity to teach geography. He was strongly supported by his Professor, J. Ainsworth Davies, in whose Department Fleure was at the time teaching zoology, botany and geology. Ainsworth Davies saw clearly that the full development of geography at Aberystwyth would require both time and financial resources but had no doubt that Fleure was the man to develop it. As a result of these deliberations Fleure's teaching duties were reorganized. His work in botany was given over to another Department but for the time being he continued his work in zoology and geology, while being appointed additionally to the vacant

Lectureship in Geography. Three years later Fleure became Professor of Zoology and in the same year a Chair of Geology was established with O. T. Jones as Professor. Jones had obtained a First-Class Honours degree in Geology on the same day as Fleure achieved his in Zoology in 1901. Based on this appointment there developed a close association between geography and geology for many years to come.

As the teaching of physical and human geography was in this way being consolidated in the College, Fleure worked hard to increase the student catchment area. He arranged courses for student teachers and diploma courses for others who could fit in secondment from their schools to attend such courses for one year. Summer courses similar to those based on the University of Oxford were also organized. The Oxford Summer Schools were in the charge of Professor A. J. Herbertson and it followed that the two men met on several occasions, both at Oxford and Aberystwyth, and so influenced one another's teaching and thinking on geography. In later years Fleure often spoke of the importance of this contact and of his admiration for Herbertson's wisdom and of 'treading with him on close and common ground'. This is a very important matter when attempting to analyse the work undertaken in geography at Aberystwyth when a full degree-granting department was established in 1918. The one-year Diploma courses were becoming well known and well attended. Their academic content was reaching a high standard and anticipating the full degree courses of the University at a later date.

The third and final preparatory phase at Aberystwyth followed the death of Herbertson in 1915. So closely had Fleure identified his ideas with those of Herbertson that when it came to the selection of a successor to Herbertson as Honorary Secretary of the Geographical Association, Sir Halford Mackinder invited him to take the office. A little later Fleure was appointed Honorary Editor of the Association's journal – a post he was to hold for thirty years. This extended Fleure's interest in school geography and drew him even closer to Herbertson's philosophy of the subject, and all at a time just prior to his achieving his chief desire – that of becoming a Professor of Geography. In 1917 the Misses Davies of Gregynog donated a sum of £20,000 for educational purposes at Aberystwyth to be directed towards those aspects of education that dealt with international affairs and understanding – studies which many felt would follow the War. They knew of Fleure's work in geography and wished that something should be done to further work and study in this direction. The University of Wales had already agreed to accept geography as an honours degree course, and so Fleure was desperately anxious to exchange his chair in zoology for a new one to be based on the Gregynog gift to the College. This was agreed to in principle. At first it was thought that the appointment should be on a temporary basis to see how the situation developed, but finally the donors were asked to endow the proposed Chair in perpetuity. This was agreed to and on 29 May 1918 Fleure was appointed to the new Chair which he himself, in conjunction with the Misses Davies, decided should be entitled 'the Gregynog Chair of Geography and Anthropology', thus bringing together the two main lines of development that had been uppermost in his mind during the whole of the preparatory period. He explained to the University that the new department was not to be a Department of Geography and Anthropology considered separately but that he wished it to be 'a Department of Human Geography in the widest sense of the term'. The new department was the first in Britain that enrolled students to read for initial and higher degrees in geography in both the Faculties of Arts and Science. The Universities of Oxford, Cambridge and London had been anticipated, and even the University of Liverpool, which had established a Chair of Geography a few months before Aberystwyth, confined its work to the Faculty of Arts.

It is appropriate now to turn to a brief examination of the work and teaching of the Department in its earliest years and to see how the ideas of Fleure and Herbertson interlocked in the formation of the syllabus, the overall planning of which, in the period 1918–20, was an obvious task for Fleure. Briefly, his academic outlook rested on the fact that in the mid-nineteenth century the work of Charles Darwin had shown the unity and order of animate Nature and that the evolutionary concepts he propounded clearly demonstrated that Man was a part of Nature. Furthermore, Darwin's methods and principles had revolutionized scientific thinking. Fleure had confessed that he was particularly attracted to Darwin's ideas when still a schoolboy and that he felt a desire to study his methods and principles further; hence his decision to take a degree in zoology when he first came to Aberystwyth. As his studies progressed he realized more and more that Darwin's evolutionary concepts had demonstrated clearly the continuous interaction of Man and all living things with their environment. In the case of Man's evolution it was not only the study of Man in his relationship to his environment but also the study of environments in their relationship with Man. Geography, he maintained, set the environmental stage for successive scenes in the drama of human experience in the various regions of the world. This, in turn, involved the study of the development and configuration of life on the Earth's surface which was enshrined in the record of history and archaeology. In his presentation of the relationship of men and their environments Fleure had little time for those who attempted to devise 'laws' in human geography, and such well-known terms as 'determinism' and 'possibilism' did not enter into his writings or teaching. Likewise, he had little time for those who maintained that we have as much as we can do to study the actual world as it exists at present. More important, Fleure maintained, was to see how the present world had come to be what it is – that is, he adopted an evolutionary approach: 'only in this way can we get beyond gazetteer description into the region of under-standing of the present world ... In such an approach we meet, as geographers, with anthropologists, historians and pre-historians.' Here, then, is a trilogy: Anthropology, History and Geography, 'to be torn asunder only with severe loss of truth'. This is, to use modern terminology, the basic Fleureine 'model for geographers'.

Herbertson differed from Fleure primarily in having been brought up in the physical, rather than the biological, sciences, and his early work in both Edinburgh and Oxford was deeply entrenched in this direction. Very early in his career he interested himself in the distribution of natural phenomena, such as rain and wind, temperature and air pressure. From Edinburgh Herbertson went to the Universities of Freiburg and Paris and then on to Montpellier where Flahault was doing pioneer work in plant associations. On returning to Scotland he worked on the famous Meteorological Atlas prepared by Bartholomew and Buchan and finally moved to Oxford to help with geographical studies there. He continued his research into rainfall distributions, particularly that of mean monthly rainfall, over the land surfaces of the globe. This is a study of the distribu-tion of a phenomenon which is a short-term variable with a long-term variability on the average. This was the type of phenomenon on the Earth's surface which interested him greatly as opposed to the immensely long-term changes of uplift, sea-levels and other geomorphological matters. Geomorphological matters do not loom large in any of Herbertson's sylla-buses for geography, neither did they form an important part of his lec-tures to the various Summer Schools that were held regularly at Oxford or in similar courses that he gave as a visitor to Aberystwyth.

His teaching of systematic geography was eclipsed after 1905 by the publication in the *Geographical Journal* (25: 300–12) of his well-known paper, 'The major natural regions' of the globe, though he based his scheme of divisions mainly on climate. First of all, temperature, pressure and rainfall were all considered, though later, after his stay at Montpellier, he was led to look upon the distribution of natural vegetation as a sort of synthetic result of the varied climatic influences. All this is known to have been discussed with Fleure in Aberystwyth in 1915 shortly before

Herbertson's death. It is possible that, because Herbertson had become more and more of a biologist, the two men found themselves 'walking together along the same path'. Fleure had also concluded that Herbertson had, indeed, been led to the Darwinian point of view, albeit by a different road (see 'The later developments in Herbertson's thought: a study in the application of Darwin's ideas', the Herbertson Memorial Lecture for the Geographical Association given by Fleure in Tenby in 1952 and published in *Geography* (37 (1952): 97–103). The impact of the paper on the natural regions of the globe, and subsequent discussions with the author, left a great impression on Fleure and stimulated him to write that Herbertson's paper had become 'an essential feature of geographical thought'.

Fleure, however, proceeded to argue that a strong case could be made out for regions with special reference to man and his work, an approach he developed in his well-known paper on 'Human regions', published in the *Scottish Geographical Magazine* (35 (1919): 94–105). Before discussing Herbertson's great emphasis on regional geography, both in his text-books and scientific papers, we should not overlook the fact that he always stressed that regions should be studied comparatively – 'the sober geographer is concerned to analyse resemblances and differences between various lands and their peoples to see how far the conditions in one correspond with those in another.' Fleure accepted this comparative approach to the study of regions as it was based on an evolutionary idea, and he even stressed the importance 'of parallels in evolution' long before many biologists did. It was in this way he could emphasize that there often arose different kinds of human response in regions possessing analogous physical environments. With this background and the strong Herbertsonian traditions, regional geography loomed large in the first degree syllabus at Aberystwyth.

This first syllabus may now be examined in some detail, particularly as it reflects the progressive thought and discussions about Geography over the long preparatory period. There were fourteen courses in all, ten lecture courses and four practical classes. The Honours scheme involved a three-year course of training.

In the first year students followed four theoretical courses and one practical class. In addition all first-year students were requested to attend a first-year course in geomorphology given in the Department of Geology. This was to make up an obvious deficiency on the physical side – a legacy of the Herbertsonian tradition. The first-year courses included a study of the distribution of land and water; the relief of the major continental masses; long and short movements of the lithosphere and hydrosphere; the atmosphere and its movements; and climates of the globe. A second course at first year, presented by the Professor, was a world survey of the major natural vegetation regions ranging from the equatorial forests to the tundras. The syllabus stated that the treatment would make special reference to climatic factors, vegetational characteristics, human activities and social organizations. Here the influence of Herbertson is clearly marked while Fleure had inserted 'human activities'. The third and fourth courses were regional which followed the general pattern of the great French regional monographs – structure, orography, drainage, climate and weather, agricultural and manufacturing areas, industries, settlements and lines of communication. The regions selected were the British Isles and America, north of Mexico. The practical class dealt with map work: Ordnance Survey maps, map calculations, preparation of maps and block diagrams, and a study of maps associated with Daily Weather Reports. It is worth noting at this point that Fleure and his colleagues placed great importance on map studies during the three-year course, emphasizing that the map was the geographer's special means of expression.

The second-year work included a course on the historical geography of commerce; three regional courses; and two practical classes. Historical geography of commerce included evidences of trade and courses on prehistoric times, early Mediterranean trade, Hanseatic Leagues, medieval trade with the Middle East, the voyages of discovery, the development of ships and the relationship to trade routes and materials, the Industrial Revolution, the coming of the railways, air transport and routes. It is obvious that this course had a strong evolutionary basis with the study of early ships, railway engines and aircraft treated as 'geological fossils'. All members of staff participated in the course. The three regional courses were treated much as those given in the first year. One dealt with the 'lands of the Romance languages' and contained material on the prehistory and physical anthropology of those countries. The second course dealt with the 'three southern continents', treated comparatively. Here Herbertsonian influence was clearly marked as indeed it was in the third course entitled 'The Monsoon Lands' with its climatic regional base. The practical classes at this stage concentrated on the field- and office-work involved in various topographical surveying processes and the construction of simple map projections, while atlas maps were analysed with particular reference to structural geography.

The final Honours year contained two further regional courses. The first was an advanced study of the Homeland entitled 'the British archipelago', concentrating on its world position and economic geography; the other was entitled 'the lands of ancient civilizations'. In practice this worked out as a study of Egypt and the Nile valley; the Tigris and Euphrates plain; and the Indus valley. Here was an ideal opportunity for the geographer to become acquainted with the contributions of both archaeology and proto-history in illuminating the evolution of human society in a specialized physical environment and for the three riverine areas to be treated comparatively. Here was something in the Fleure–Herbertson tradition with an evolutionary emphasis that was characteristically Fleure. In later years, between 1927 and 1936, Fleure published much of the lecture material, greatly expanded, in the well-known series which he wrote with H. J. E. Peake, *Corridors of Time*. The titles of the successive volumes vividly illustrate the evolutionary approach: *Apes and Men, Hunters and Artists, Peasants and Potters, The Steppe and the Sown, The Way of the Sea*, and so on.

The second main Honours course apart from the regional courses was entitled 'the races of man'. It dealt with the origins of mankind and the distinguishing characteristics of the human races in skin, hair, eyes, skull, limbs, stature and other physical attributes. The distribution of these characteristics was studied in relation to climate in particular, and their permutations and combinations in different peoples loomed especially large in the course. Fleure maintained throughout that we cannot be students of human geography without knowing all that can be known about humanity, physically and culturally, as well as geographically. The map work in the Honours practical classes dealt with advanced survey work in the field and with the elements of geodesy and field astronomy. Particular attention was devoted to map projections and graticules of special form such as interrupted networks and their underlying principles.

It is useful at this stage to review the work of the Fleure years, 1918–30, by a brief examination of the thesis topics accepted by the University for higher degrees from students of the Department during the period. Altogether thirty-seven M.A. or M.Sc. degrees by research were awarded. Six dealt with topics in historical geography and another six in economic geography. Five presented topics in agricultural geography, dealing mainly with land utilization studies, and five others were in human geography, with topics in population statistics and settlement patterns. By contrast, five theses were submitted on purely archaeological topics, four on studies in social anthropology and two in physical anthropology. Four postgraduate theses were submitted in physical geography including vegetation studies, historical cartography and map projections. If the four classed as physical geography are excluded, there remain thirty-three theses, of which twenty-two are on topics in human geography and eleven on archaeological and anthropological subjects. Whatever else these figures may show, it is fair to say that in the final reckoning during the Fleure period most of the more able, serious students were interested in geographical studies rather than in anthropology or prehistoric archaeology as such.

In 1930 Professor Fleure accepted an invitation from the Victoria University of Manchester to become the first holder of a newly created Chair of Geography within a department which at that date had not grown sufficiently to the point of establishing Honours courses. He was attracted to Manchester by the opportunity it offered of establishing the subject in a university located in a major metropolitan setting. His desire to set up full Honours courses in every university in the land was well known and he felt, in consequence, that the invitation from Manchester could not be set aside. His first objective was, therefore, to found an Honours school and this was achieved in two stages, first in the Faculty of Arts only, and a year later in the Faculty of Science. The Honours school in the latter Faculty appeared to follow closely the Aberystwyth scheme whose students studied geography and anthropology and were required to complete a Part I course in geology before embarking on their Honours work. In this way it can be said that, by and large, Fleure carried to Manchester the concept of Geography that had developed at Aberystwyth. One thing that appeared to emerge was that the response to Fleure's teaching of physical anthropology was not received with the enthusiasm that it had been received in Wales – only a few research students wanted to 'measure heads' and record other anthropometric data in the Pennine hinterland. At the outset also Fleure participated less in standardized regional courses and concentrated more and more on systematic studies, including in particular his increasing interest in, and contributions to, urban geography. His interests in the evolution of early human societies remained unabated and much research work in the Department concentrated on an exhaustive study of the Megalithic culture. He insisted that it was necessary for students mapping Megalithic remains to know all that could be known about the culture at first hand. This naturally led not only to wide reading but also to extensive field-work in Ireland and in the Isle of Man where research students from the Department of Geography under Fleure's personal guidance and direction became engaged in actual archaeological restoration work on Megalithic tombs.

Fleure's interests in the teaching of Geography remained in the forefront of his work. He continued to be both Honorary Secretary of the Geographical Association and Editor of its journal. The office and library of the Association were in many ways an adjunct of the Department in Manchester as they had been at Aberystwyth. Furthermore, it must be remembered that he had taken with him to his new environment many of the philosophical ideas that he cherished at Aberystwyth, especially those regarding the significance of geography teaching in schools and universities and its importance in helping to create good citizens and strengthening the cause of world peace and international understanding. During his time in Manchester the cultural environment of the City would appear to have helped the presentation of such ideas. His stay at the University coincided with the stresses and strains of the Second World War and the rise of Nazi Germany and the humiliation of France. The latter, naturally, concerned him greatly. The result was that these matters loomed large in his geographical thinking and formed the content of his seminar work and discussions with students, and also of his lectures both in the University and outside its walls. This type of work led to an interesting development in his teaching of regional geography. At the end of his period in Manchester the selected regions for study veered from regions with geographical appellations such as the 'Mediterranean lands' or the 'Monsoon lands' to the use of political units – nation states such as France or Belgium or even the super-states of the USA or the USSR. Here Fleure was drawing nearer to many other regional geographers (particularly those in the University of London) who realized that nation states of this type were the most important 'regions' in the modern world. All these changes of emphasis did not alter Fleure's basic model of teaching, worked out at Aberystwyth, resting on his famous trilogy – anthropology, archaeology, geography. He remained in Manchester until his deferred retirement in 1944 towards the end of the Second World War.

Meanwhile at Aberystwyth Cyril Daryll Forde had been appointed to succeed Fleure as Professor of Geography and Anthropology. He differed from his predecessor in that he had graduated in Geography and had taught the subject for a short while as an Assistant Lecturer in the University of London. Between 1923 and 1928 he had done fieldwork research on the prehistory of Brittany. It is significant in the present context that his doctoral thesis on the prehistoric geography of Britain had a distinctive geographical bias. Subsequently, he spent two years at the University of California studying the pre-Columban civilization of America. He had, therefore, acquired a scientific training in prehistoric archaeology, social anthropology and geography, and began immediately putting them into practice as he recast the teaching syllabus at Aberystwyth. He realized, however, at the outset the Department's limitations in providing an adequate training in physical geography and in his second session he persuaded the College Senate to approve a scheme of instruction in physical geography for the first and second years. The scheme was drafted in co-operation with the Professor of Geology and was to be given in the Department of Geology. The new professor found it difficult to deal adequately with the Honours course in physical anthropology and soon handed it over to another member of staff who had been trained under Professor Fleure. The Honours examination in 1932 was composed of six written papers. One dealt with the races of Man, a second with early civilizations, and a third with regional work on North America under the Professor's care. The fourth paper dealt with the regional geography of France, as the Professor felt that regional geography should be studied much on the lines of the classical traditions of the Sorbonne; an extra member of staff (who had studied at the Sorbonne) was appointed for this course. There was, in addition, another regional paper on the Southern Continents, and a General Paper where a variety of topics were listed, including some very general ones on historical geography. There followed the usual practical work.

Behind this extensive teaching programme Professor Forde and his colleagues continued with their individual research. Forde, besides writing an extremely important and searching paper on the content of human geography ('Historical geography, history and sociology', published in the Scottish Geographical Magazine (55: 217)), turned to serious research in prehistoric archaeology, much in the same way as Fleure had turned to fieldwork on selected Megalithic tombs. With the assistance of senior students, he devoted three summer vacations to excavating the nearby Iron Age B hill-fort at Pen Dinas, producing valuable scientific results. Throughout his teaching, however, it was abundantly clear that he was primarily interested in the explanation and distribution of human activities on the Earth's surface in a broad general way, following closely Fleure's example. He differed from his predecessor, however, in attempting to make the study of the relationship of Man and his Environment more precise. Fleure disliked 'laws' in human geography while Forde participated enthusiastically in discussions associated with 'determinism' and 'possibilism', falling back on Lucien Febvre's famous dictum – 'there are nowhere necessities but everywhere possibilities, and Man as the master of these possibilities is the judge of their use'. Forde maintained that such ideas lay behind his well-known text-book, Habitat, Economy and Society: a geographical introduction to ethnography, published in 1934. This book was widely used by teachers of geography in schools and colleges, very largely for the detail it contained on primitive societies;

yet it was felt that many of the pupils who used it often failed to see the complexity of primitive human cultures, and to appreciate how such complexity is nowhere more manifest than in the study of the adaptation of human societies to their environments.

Two matters of importance stemmed from this book that influenced the course of geographical studies at Aberystwyth. On the one hand, Forde became more and more certain that progress lay in further field studies among pre-industrial societies where conditions of adaptation to both the physical and cultural environment were simpler than those associated with complex industrial communities. With these ideas in mind Forde proceeded in 1935 to carry out an intensive field study in eastern Nigeria among people practising a tropical subsistence economy. During his absence on field studies of this nature his abler students, and others, wanted him to direct his attention to the need for similar research nearer home. This request led to research work such as that of A. D. Rees, whose study of a village community in North Wales resulted in his well known work *Life in a Welsh Countryside.* Although it was not published until 1950 the author readily acknowledged Forde's influence, and the book, in turn, stimulated further studies of this kind. While this research work was going on there grew up an unmistakable opinion among the majority of students and teachers in other departments that university geography should be taught on what they considered to be orthodox lines. If we make an analysis of the higher-degree theses submitted during Professor Forde's regime we can see where the interests of the senior students lay. Archaeological and anthropological topics were avoided to a greater extent even than in Fleure's day. Of the thirty-one theses accepted by the university not a single one dealt with a topic in physical anthropology, and only two were concerned with archaeological subjects and three with topics in social anthropology. On the other hand, there were eleven theses in human geography, five in physical geography, four in agricultural geography, and three each in historical and economic geography. Three theses were accepted on miscellaneous topics which do not fall into the above categories.

In 1941 after his return from a further period of fieldwork in Nigeria Forde was given more leave of absence for war duties on the staff of the Foreign Office Research Department, first in Oxford and later in London. As the War drew to its close in 1945 Forde resigned his post at Aberystwyth on his election to the Chair of Anthropology (not Geography) in the University of London at University College London. On the receipt of Forde's resignation the College Council invited the Senate to prepare a report on the scope and nature of the work to be undertaken within the Department of Geography and Anthropology with special reference to the duties of a newly appointed Professor. This report summarizes succinctly the general situation as it then existed. It stated:

> In considering the needs of the Department it should be remembered that special attention has always been paid to human geography in its widest sense. The position is indicated by the title of the Chair though it is worth emphasizing that geography here takes precedence over anthropology, as it certainly does in teaching and research. The Senate is of the opinion that the emphasis should remain on human geography rather than upon anthropology in the specialist sense, and that this should be borne in mind in considering the appointment of a Professor.

It was in this atmosphere that the third Professor of Geography and Anthropology at Aberystwyth, E. G. Bowen, was appointed, and although his work and teaching while occupying the Chair lie outside the scope of this volume, it is important to note that it was during his period of service that the two main problems that the Department had inherited from the outset were solved. Anthropology was omitted from the title of the Chair, after consulting legal opinion regarding the original bequest, and the College created a new and fully staffed Department of Sociology and Social Anthropology under a new professor. By 1965 Bowen had also succeeded in persuading the College authorities that a second Professor was required in the Department of Geography and that he should be designated 'Professor of Physical Geography', apart from any other administrative title he might hold. At the same time it was agreed that the older Chair should retain the name of the family home of the donors and be designated 'the Gregynog Chair of Human Geography'. Professor Clarence Kidson was appointed to the former and Professor Harold Carter to the latter.

While the Aberystwyth Department grew along these lines, there was also a development of geographical studies in the University College of Swansea, a sister College in the Federal University of Wales. From its foundation as a University College in 1920, Swansea had in mind the need to concern itself with the teaching of Geography. The College had appointed at the outset a distinguished Professor of Geology, A. E. Trueman, who had written several papers on geographical topics, including some on population distribution in South Wales. In 1931 D. Trevor Williams, a former student at Aberystwyth, with a specialist interest in historical geography, was appointed as Lecturer in Geography attached to the Department of Geology. In 1934 T. Neville George succeeded A. E. Trueman with the title of Professor of Geology and Geography, with D. T. Williams retaining the lectureship in geography under the Professor of

Geology and Geography. Just prior to the Second World War an independent lectureship in Geography, with Williams in charge, was created. In 1946 B. H. Farmer succeeded Williams who, like many other geographers at this time, joined the Ministry of Town and Country Planning. The personnel changed yet again in the early fifties with Dr Duncan Leitch becoming Professor of Geology and J. Oliver being appointed Lecturer in charge of the Department of Geography. The session 1954–5 saw a major development at Swansea when W. G. V. Balchin, a Cambridge graduate working in physical geography, became the first Professor of Geography and Head of a new independent Department of Geography.

We must revert to the Fleure period at Aberystwyth to note that several senior students obtained university teaching posts. Although they naturally tended to present the Fleure viewpoint, and some did research on Fleure-like topics, it must be noted that almost invariably they joined departments, which, though youthful themselves, still had developed a character of their own. This applied to D. T. Williams, who went to the University of Exeter, and to others who were appointed at Bristol, Manchester and Leeds. In such positions these teachers had no executive authority and could not plan the curriculum of their departments as a whole. This means that the spread of Fleure's teaching and viewpoint was carried in the first instance by individual teachers in this way. There was, however, one important exception to this situation – the appointment of Emyr Estyn Evans to The Queen's University, Belfast, in 1928 as an independent lecturer in Geography and Head of Department. Evans had the opportunity here of building up an entire School of Geography from scratch, as it were, and in this sense he can be ranked with the pioneers of the inter-war years. He graduated in Geography with First-Class Honours in 1925 and proceeded direct to Belfast, three years later, remaining there for forty years and becoming the first holder of the Chair of Geography in that University in 1945. Full Honours courses in the subject were established at the same time in both Faculties.

Because Evans not only left Aberystwyth very early in his career and went straight to Belfast, it is understandable that he should model the new Department very much on Fleure-like lines. It must also be remembered that he was entering virgin territory as far as the teaching of geography was concerned. He received, therefore, little help from senior schools and so had to start literally 'from the beginning' with his students. As time went on, however, this proved to be a great advantage in that by 1945 Evans had laid a broad, sound, orthodox foundation for the subject apart from his own original contributions to geographical studies in his own specialized fields. This aspect of geographical teaching in Belfast is often overlooked because of the brilliance of Professor Evans's own contributions to research.

Throughout Estyn Evans followed Fleure and other geographers in emphasizing that the core of geographical studies was the study of the relationship of Man to his Environment, both physical and cultural. It followed naturally that he organized courses dealing with the races of Man and physical anthropology generally. He did not, however, seek to make serious anthropometric studies in the field as Fleure had done in Wales although he was fortunate in Belfast in that there was a Medical Faculty at hand and a Professor of Human Anatomy ready to co-operate. More important in his teaching and research was his interest in the great heritage to be found in the Irish environment – an environment unique in many ways, seeming to lie outside history. Evans and his students began this study by concentrating on rural settlements and making many original contributions to this interesting geographical field. This work led on in turn to the Department's interests in folk life and culture. The Professor realized at the outset that no understanding of the Irish heritage was possible without digging deep in the past. Like Fleure he was fascinated by the evidence in the Irish landscape of the remains of the Megalithic culture of five millenia ago; and so he devoted a great deal of his interests, and encouraged his students to do likewise, to surveying and excavating these Megalithic tombs under expert supervision.

Regional geography and physical geography were certainly not forgotten in Belfast. Evans took with him the teaching of regional geography as developed at Aberystwyth with the regional geography of the major continental areas forming the most obvious line of approach. He had, however, been influenced by Fleure's admiration of the French school of regional geographers at the Sorbonne in the days of Vidal de la Blache, and by the time he arrived at the Queen's University he had developed a deep interest in France and its culture. He was fluent in French and used various parts of France for field excursions with his students and finally wrote an interesting work on that country. Even more important is the fact that he was able to show that by welding Habitat, Heritage and Geography it was possible to produce an excellent regional study selected from the Irish landscape. His *Mourne Country* (1951) ranks in concept and treatment with the great monographs of the Sorbonne school. It is significant that this type of regional treatment impressed him more than all Fleure's writings on regional themes. One was a scientist and the other an artist. It is, therefore, not surprising to note that when Professor Evans retired from his Chair of Geography he was presented with four *Festschrifts*: in Ethnology, Archaeology, Cultural Geography, and Regional Geography.* These summarize neatly the interests of the Professor and his Department. They were guidelines of 'the model' on which the Professor worked.

It may be useful by way of conclusion to review the growth and expansion of the teaching of the Fleure School of Geography from its beginning at Aberystwyth to the death of Fleure in 1969. It can be safely said to have spread widely from Wales at the hands of the Founder-Teacher himself and his successors, not only throughout Britain, but, indeed, throughout the English-speaking world overseas, and other countries as well during the lifetime of the Founder.

Three major factors stand out in any attempt to assess the rapid growth and widespread influence of Fleure's teaching at Aberystwyth. First of all, everyone recognized the personal factor. Fleure's teaching in practice was more like that of a professor in a medieval university, who did not of necessity have a rigid lecture or teaching programme, but was a professor who gathered a following of keen students around him and taught them either individually, or in small groups. Indeed, one of Fleure's contemporaries (also a professor of geography) put the matter succinctly: 'there is no Department of Geography at Aberystwyth, only a Personality'. This almost medieval picture is certainly very apt, for those who have spread Fleure's teaching over the world came from small groups of honours and research students who consulted him most frequently concerning their theses or research work. The second factor, as far as the bulk of Fleure's teaching was concerned, which attracted the mass of the undergraduates, was the way in which he was able to integrate what may be called 'the straightforward geography' of Herbertson (i.e. lectures on relief, structure, climatic and natural vegetation zones, population distributions, etc.) with his own more specialized interests in physical anthropology and prehistoric archaeology. Fleure as a teacher acquired his 'straightforward geography' direct from Herbertson while at Aberystwyth, and he saw in it the possibility of its providing an obvious and ideal setting for his specialized interests concerning the study of Man. It was Fleure and his colleagues presenting this 'straightforward geography' that provided the daily fare for the mass of undergraduates, who were, at this time, almost all destined to be secondary-school teachers.

* The volumes are: *Irish geographical studies in honour of E. Estyn Evans*, ed. N. Stephens and R. E. Glasscock (1970); *Man and his Habitat: essays presented to Emyr Estyn Evans*, ed. R. H. Buchanan, E. Jones and D. McCourt (1971); *Studies in folklife presented to Emyr Estyn Evans*, ed. D. McCourt and Alan Gailey, *Ulster Folklife*, vols. 15–16 (1970); *Papers presented to Oliver Davies and Estyn Evans*, ed. D. M. Waterman, *Ulster Journal of Archaeology*, 3rd series, vol. 33 (1970). (Oliver Davies has been Professor Evans' long-time friend and collaborator.)

Herein lies the third factor that accounts for the rapid spread and reputation of the Fleure School of Geography in its initial stages. As already noted, Fleure had taken over from Herbertson in 1915 not only the secretaryship of the Geographical Association but also the editorship of its quarterly journal, then called the *Geographical Teacher* (and later *Geography*). His small secretarial staff and those responsible for the publication of the journal were all together in the same building as the Department of Geography, so that Fleure was virtually Director, Secretary and Editor of the Association's journal all combined. This was a very powerful instrument for spreading Fleure's teaching as the *Geographical Teacher* went to nearly every teacher in the secondary schools of Britain who taught geography at Higher (later Advanced) level. No other professor of geography had such a valuable organization almost directly under his immediate control. At the same time Fleure exerted considerable pressure with other members of the Association on the then Ministry of Education to increase facilities for the teaching of geography in schools and for its examination by the major Examining Boards. In this way, Fleure's name became known throughout the land among teachers of geography, while at the same time he was, in his capacity as a senior professor, closely involved in pressing for the establishment of chairs of geography in all the British Universities; and he himself was often closely concerned with professorial appointments. The ground, therefore, had been well sown before Fleure left Aberystwyth for Manchester.

The appointment of Professor Fleure's successor at Aberystwyth naturally presented a crisis in the affairs of the Department. Although it was well known that Daryll Forde was not Professor Fleure's choice, the appointment was, however, considered at the time as an ideal one. Even as a young man Daryll Forde was admirably qualified academically, both as a geographer and as an anthropologist. Nevertheless, other considerations, including the somewhat Victorian academic climate at Aberystwyth at the time, made it difficult for him to continue the Fleure tradition as a matter of course. Furthermore, he had imbibed very liberally of American culture, both socially and academically, during his stay in California and was returning to Britain some years before the advent of American troops in large numbers during the Second World War period. Thus coming to Aberystwyth in 1930 was for him distinctly exotic. On the academic side, for example, Forde was convinced that the average undergraduates whom he had met did very little reading, even of text-books, but relied on repeating lecture material. He, therefore, adopted the then current technique in American universities for dealing with this matter by instituting one-word 'quiz tests'. His students were horrified and ear-marked this technique as more fitting to the kindergarten than to the university. They revolted instantly and the professor was obliged to drop the practice abruptly.

There were also other difficulties. Forde's academic approach to the subject was definitely more precise and scientific than that of his predecessor especially in his presentation of the more specialized studies of prehistoric archaeology and physical anthropology. Young undergraduates tended to be overburdened, and to lose interest, and wondered whether these subjects were in any way 'geographical' as they understood these things. So it can be said that the outlook and approach (but certainly not the scholarship) of Daryll Forde presented a real break with the Fleure traditions. It was in this atmosphere that the College Senate felt the necessity of making its points to the College Council, as noted above, for consideration when the time came to appoint a successor to Forde. Highly qualified candidates were turned aside in favour of a candidate who was considered to be directly in the Fleure tradition, and who had developed the closest links with schools and colleagues in both England and Wales. As has been pointed out, E. G. Bowen, when in office, endeavoured to keep much of the Fleure pattern but by the strengthening of the teaching of regional geography and minimizing the teaching of physical anthropology, and by pressing for the creation of a second Chair of Physical Geography he tended to bring geography teaching at Aberystwyth more in line with that in the University of London than had hitherto been the case. This is a matter that is not always fully appreciated in university geographical circles, where Fleure had left so great an impression that it became customary to think of Forde's successor in the Aberystwyth Chair in 1946 as somewhat reminiscent of 1660 in English History – the Restoration of the Monarchy! In fact, the overall situation appears to indicate that the Fleure tradition has been more fully maintained in The Queen's University in Belfast than in either Manchester or Aberystwyth.

REFERENCES

E. E. Evans (1951), *Mourne Country.*
H. J. Fleure (1919), 'Human Regions', *Scottish Geographical Magazine*, 35, 94–105.
(1952), 'The later developments in Herbertson's thought: a study in the application of Darwin's ideas', *Geography*, 37, 97–103.
H. J. Fleure and T. C. James (1916), 'The geographical distribution of anthropological types in Wales', *Journal of the Royal Anthropological Institute*, 46, 35–153.

C. D. Forde (1939), 'Historical geography, history and sociology', *Scottish Geographical Magazine*, 55, 217–35
   (1934), *Habitat, Economy and Society: a geographical introduction to ethnography.*
A. J. Herbertson (1905), 'The major natural regions', *Geographical Journal*, 25, 300–12.
H. J. E. Peake and H. J. Fleure (1927–56), *The Corridors of Time.* The complete series is: *Apes and Men* (1927); *Hunters and Artists* (1927); *Peasants and Potters* (1927); *Priests and Kings* (1927); *The Steppe and the Sown* (1928); *The Way of the Sea* (1929); *Merchant Venturers in Bronze* (1931); *The Horse and the Sword* (1933); *The Law and the Prophets* (1936); *Times and Places* (1956).

4 Geography at Birkbeck College, University of London, with particular reference to J. F. Unstead and E. G. R. Taylor

EILA M. J. CAMPBELL*

For all but one of the years 1918–45, the geography department at Birkbeck College was 'guided' successively by J. F. Unstead and Eva G. R. Taylor. Unstead was appointed lecturer in geography at Birkbeck College in 1909 following George G. Chisholm (at Birkbeck, 1895–1908) and L. W. Lyde (at Birkbeck, 1908–09); the latter lectured at Birkbeck in a part-time capacity while also occupying the chair of geography at University College London, to which he had been appointed in 1903. Shortly after Birkbeck College became a constituent school of the University of London in 1920, Unstead was appointed to the newly created chair of geography tenable at the College.[1] Ten years later, at the early age of fifty-five, he resigned from his post 'in order', he has gone on record as saying, 'to read and think, to travel and write'. He was succeeded by Dr Eva Taylor who was appointed to the chair in open competition. She had studied in Oxford under A. J. Herbertson between 1906 and 1908 for the Certificate of Regional Geography and the Diploma of Geography, both of which she obtained with marks of distinction. She also served from 1908–10 as a research assistant to Herbertson. She used to compile and draw his wall maps for schools and was paid privately by him. She first joined the staff of Birkbeck College in 1921, having previously lectured in a part-time capacity at East London College, later Queen Mary College. She had also lectured in geography and educational

* Eila Muriel Joice Campbell (b. 31 December 1915) trained as a teacher at Brighton Diocesan Training College for Teachers, 1934–36. She graduated in geography (honours) in the University of London in 1941. She was appointed to an assistant lectureship in geography at Birkbeck College in 1945 and was later lecturer and, from 1963, reader in geography. In 1970 she was appointed to the Chair of Geography, a post which she held until her retirement in 1981, when she was made an Emeritus Professor of the University of London.

method at Clapham Training College for Teachers and at the Froebel Educational Institute. Among the applicants for the chair was S. W. Wooldridge who was to be appointed Eva Taylor's successor when she retired in 1944.

Others who held full-time posts in geography at Birkbeck College between 1918 and 1945 were H. A. Matthews, H. C. K. Henderson and A. C. O'Dell. H. A. Matthews played a significant part (with R. O. Buchanan and S. W. Wooldridge) in the foundation of the Institute of British Geographers and was its first Assistant Secretary. O'Dell became the first incumbent of the chair of geography established at the University of Aberdeen in 1951 (six years earlier he had been appointed lecturer in geography and head of department of geography at Aberdeen). Henderson was appointed to a second chair of geography at Birkbeck College in 1965, where W. G. East had succeeded S. W. Wooldridge in 1947 when the latter returned to King's College London to occupy a newly created chair of geography tenable at the College.

Throughout the period under review (1918–45) British geographers were concerned not only with regional description (Unstead 1932) but also with the theoretical principles of regional division. In this they were not alone. Among British geographers, Unstead played a leading role in trying to make regional description more scientific. Geographers in other countries were also concerned with the identification and classification of regions – both great and small. Among those working on similar lines to Unstead were two German geographers, Otto Maull (1936) and Peter Heinrich Schmidt (1937).

Unstead (1926) first presented his scheme for a hierarchy of regions in a paper on Spain in the *Scottish Geographical Magazine*. Seven years later he carried the ideas set out in that paper further in his Herbertson Memorial Lecture delivered to the Geographical Association (Unstead 1933). In the second paper, he suggested the term 'stow' for the smallest unit of any regional division and that of 'tract' for a contiguous group of interrelated 'stows'. Thus he 'categorized' the South Downs of England as a 'tract' consisting of a number of transverse valley floor 'stows' (e.g. the Adur and Arun valleys) separated by interfluvial plateau 'stows'. The physiographic unit of south-east England itself was a group of 'tracts' (scarps and vales). In this paper he also discussed the work of the German geographer Siegfried Passarge (1866–1958)[2] which had clearly influenced his own work. Unstead's presentation in 1933 of his system of regional geography led to the Geographical Association appointing a committee to look into the classification of regions. Four years later the committee, consisting of Unstead, J. L. Myres (Professor of Ancient History at

Oxford), P. M. Roxby and L. D. Stamp, published its report (Geographical Association 1937; Dickinson 1976).

Unstead had first become interested in the classification of regions after hearing a lecture on the subject by A. J. Herbertson while attending (as an observer, on the suggestion of H. J. Mackinder) the biennial summer vacation course for teachers of geography in Oxford in 1904 (where he also met W. M. Davis, Roxby and Eva Taylor). The following year, he published the substance of his lecture in a paper entitled 'The major natural regions: an essay in systematic geography' in the *Geographical Journal* (Herbertson 1905). During the summer school Unstead was also introduced to the concept of the 'pays' as presented by Vidal de la Blache in his *Tableau de la géographie de la France* (1903); Herbertson himself is on record as being impressed by this work. Unstead first used Herbertson's scheme of major natural regions, albeit in a modified form, in a textbook entitled *General and Regional Geography for Students* which he published jointly with Eva Taylor (Unstead and Taylor 1910). It is of interest that Eva Taylor also heard Herbertson's lecture in 1904; at the time she was attending a summer vacation course in Oxford organized by the University's Department of Education. She was a first-class honours graduate in chemistry of the University of London and was in her first post as a teacher of the subject at a convent school in Burton upon Trent. In their pioneer textbook, which Eva Taylor always maintained they drafted on the steps of the British Museum, Unstead and Taylor included maps showing the natural regions of each continent. In the accompanying text, they emphasized that the great climatic and vegetation divisions of the world were the chief guides to the regions and that in addition an 'important distinction between plains and uplands or mountains' had to be made. Herbertson's scheme of natural regions and the modifications of Unstead and Taylor were widely studied and in turn copied and modified by other British geographers. These included Leonard Brooks, who had been appointed on a part-time basis to help Unstead at Birkbeck College in 1919, and L. D. Stamp, both of whom included maps of natural regions of the several continents in their textbooks, a practice which was continued throughout the inter-war period and even for a few years after the Second World War.

Unstead appears to have tried to establish a system of combining small regions into large ones as early as 1916 (Unstead 1916). In this he seems to have been influenced by two other articles published by Herbertson – the first entitled 'The higher units: a geographical essay', published in *Scientia*, an international journal devoted to the synthetic aspects of science (Herbertson 1913b) and the second entitled 'Types and orders of natural regions' (Herbertson 1914). Unstead pursued regional division and regional analysis relentlessly for nearly thirty years but his texts on regional geography were tedious (Unstead 1932; 1935), perhaps because they were overloaded with too many uninspiring facts. Nevertheless his contribution to the concept of the region and to methods of organizing the world into regions should not be underestimated. Richard Hartshorne recognized its value by citing Unstead's contributions in his long discussion of 'the fundamental function of geography – the understanding of the differences between different areas which requires the geographer to divide the world arbitrarily into areal parts' (Hartshorne 1939). Unstead's approach was to proceed from the smallest unit (readily recognizable on the ground) to the greatest (not easily visible in the days before remote-sensing). Unstead was diligent and industrious but he lacked the lively mind of his early collaborator and later colleague, Eva Taylor. She neither pursued the theory of regions nor helped Unstead to develop his system.

Eva Taylor was nearly forty-two years of age before she was appointed to a full-time lectureship in a university department of geography. Her published work, although considerable and very sound, all related to geography either in schools or in teacher training.[3] She was well aware that she needed to make a distinctive contribution to knowledge in at least one branch of geography and decided to investigate the history of English geographical thought, beginning in 1485. She never fully explained how she had come to select her chosen field of enquiry. She believed that scholarly writing depended on the appraisal of original source material. It is possible that she was pointed to her study of geographical thought by A. P. Newton (Rhodes Professor of Imperial History in the University of London) who was himself interested in the geographical thought of the Middle Ages (Newton 1926) and whom she would have come to know as a fellow member of the University's Board of Studies in Geography. She may also have been attracted to her subject by her early training in the natural sciences.

During the first seven years of her appointment at Birkbeck College, she spent her days beavering away in the Reading Room of the British Museum and in the Students' Room of the Public Record Office (both within easy walking distance of the College). During the evenings (six to nine o'clock) she lectured to students. Her teaching commitments like those of her contemporaries were diverse and relatively heavy.

Her chosen field of research was virtually untilled and by 1928 she had completed her work on *Tudor Geography 1485–1583* (Taylor 1930). She submitted it, together with related research papers, to the University of London for the degree of D.Sc. which was awarded to her. *Tudor*

Geography dealt with what she described in the preface to the printed volume as 'that fateful century or so during which Englishmen of all ranks were forced gradually by circumstances to think geographically as they had never done before ... Elizabeth's day saw the map and the globe as the necessary furniture of the closet of scholar, merchant, noble and adventurer alike ...'

Four years later she published *Late Tudor and Early Stuart Geography 1583–1650* (Taylor 1934). In the preface to this volume, she gave an indication of her attitude to learning:

> If the first half of the seventeenth century was the twilight that heralded the dawn of modern science, it was also the twilight that marked the passing of the golden age of unspecialised learning. How happy the day in which every man of liberal education could read, speculate and even write in whatever field or fields of knowledge he chose.

In each of the two volumes she 'attempted to depict the background of geographical thought and nautical theory that formed the setting' of the English voyages for trade and discovery. She interpreted geography in its widest sense and her trawl of relevant literature was extensive as indicated in her bibliographies. The central figure of the first volume was John Dee. The protagonists of the second were Richard Hakluyt and Samuel Purchas. Linked to her investigations into English geography – 'both practical and academic' – between 1485 and 1650 were her editions for the Hakluyt Society of Roger Barlow's *A briefe summe of geographie* (Taylor 1932a) and *The Original Writings and Correspondence of the Two Richard Hakluyts* (Taylor 1935); the latter still forms the basis of all Hakluyt studies.

During the 1930s she also gathered material for a third volume – on early Georgian geography – but she never wrote one. The non-appearance of the projected volume was due more to the fact that she became interested in the practical application of mathematics to navigation rather than to the outbreak of the Second World War and her evacuation to her cottage in the Cotswolds. In due course, she published two volumes on mathematical practitioners – *The Mathematical Practitioners of Tudor and Stuart England* (Taylor 1954) and *The Mathematical Practitioners of Hanoverian England* (Taylor 1966). The first appeared eleven years after her retirement from the chair of geography and headship of the department of geography at Birkbeck College in 1944 and the second in the year of her death when she was in her eighty-seventh year. These two volumes were in essence bio-bibliographical dictionaries of chart-makers, compilers of sailing directions and authors of navigational guides

– in other words of men who tried to serve seafarers. All the volumes in her chosen field of research were distinctive contributions to knowledge and have stood the test of time. She also published many valuable papers. Her contributions are too numerous to be discussed in detail. She was always conscious of the need to educate the general reader for whom she produced an outstandingly successful popular work on the history of navigation up to the time of Captain James Cook – *The Haven-finding Art* (Taylor 1956).

Related to her studies in the history of English geographical thought were the two chapters which she contributed to *An Historical Geography of England before A.D. 1800*, edited by H. C. Darby. These were concerned with 'Leland's England' and 'Camden's England' (Taylor 1936a; b). Eva Taylor never claimed to be an historical geographer but she gave courses at Birkbeck College on the historical geography of England and on that of the classical world. Her lectures were based on intensive reading and critical appraisal of other people's writings on the various man–land relationships in the areas and periods covered in her courses. Her lectures were always stimulating and thought-provoking. They were also illustrated by telling sketch-maps of her own design. She was interested in the role of the historical geographer and spoke at a joint meeting of geographers and historians held in London in 1932. In the printed report of this meeting, she is on record as saying: 'The application of the adjective "Historical" to the noun "Geography" strictly speaking merely carries the geographer's studies back into the past: his subject matter remains the same' (Taylor 1932b).

In 1930, as noted above, Eva Taylor was appointed by the University of London to the chair of geography tenable at Birkbeck College and to the headship of the department of geography. Thereafter, for at least seventeen years, she played a very significant role in advancing the cause of British geography. During these years, she twice served on the Council of the Royal Geographical Society (1931–5 and 1937–41). She also served for two years (1933 and 1934) on the Editorial Committee of the Institute of British Geographers. She was an active member of the Committee of Section E (Geography) of the British Association for the Advancement of Science and was elected President of Section E in 1939. As the annual meeting of the Association was interrupted by the outbreak of the Second World War, she was invited to be Section E's president at the first regular post-war annual meeting held at Dundee in 1947. At this meeting, she delivered her presidential address entitled 'Geography in war and peace'. It was a masterly review of the contribution of academic geographers to Britain's war effort (Taylor 1947).

Although Eva Taylor's primary research interest was in the history of English geographical thought and of the practical application of mathematics to navigation, she was involved in two issues of major concern to British geographers during the decade 1937 to 1947. These were the preparation of a memorandum for submission, on behalf of the Council of the Royal Geographical Society, to the Royal Commission on the Distribution of the Industrial Population (often referred to as the Barlow Commission after the name of its chairman, Sir Montague Barlow) and the need for a national atlas of Britain. The second developed from the first, and Taylor was deeply concerned with both issues.

In October 1937 the Commission asked the Royal Geographical Society to assist them by providing a 'memorandum or report on the distribution of industry and the industrial population with particular reference to geographical and atmospheric conditions.' The Society set up a small committee under the chairmanship of L. D. Stamp; the committee included Eva Taylor, G. J. H. Daysh, H. J. Fleure and Brigadier Macleod. Shortly after the Committee was set up the Chairman and Professor Fleure left for a four-month visit to India and the task of preparing the memorandum and of laying it before the Royal Commission passed to Professor Taylor, who acted as chairman of the Committee for most of the time, and Daysh. In the gathering of information on which the memorandum was based, they were assisted by a number of British geographers from all parts of England and Wales including (in alphabetical order) S. H. Beaver, K. C. Edwards, C. F. W. R. Gullick, H. C. K. Henderson, Gordon Manley, A. C. O'Dell, Wilfred Smith, E. C. Willatts and S. W. Wooldridge (the Committee asked the Council of the Royal Scottish Geographical Society to provide the evidence for Scotland and to submit their own memorandum).

Within six months, with voluntary help from university departments of geography, the Committee had drafted a memorandum of evidence and a portfolio of some forty-nine maps. Taylor regarded the maps as 'essential parts of the evidence, for they reveal relationships and suggest guiding principles which do not emerge from tables of statistics or verbal memoranda'. Stamp returned to England in time to give oral evidence with Taylor to the Commission. The Commission members found Taylor's use of the maps to elaborate the memorandum novel. With the aid of two masks, she was able to show very clearly what the memorandum described as an axial belt 'running from Greater London in the south-east to Lancashire and the West Riding in the north-west'. Within it were to be found a high degree of accessibility and therefore of attraction to industry. The shape of one of the masks resembled a coffin, and the axial belt was later often referred to as the coffin; the term was actually initiated by Sir Montague Barlow himself during Taylor's presentation of the maps to the Commission. Eva Taylor was a born teacher and her demonstration showed the value of maps in putting across complex relationships.

The Society's memorandum was printed in the *Geographical Journal* (Taylor 1938a; b) and contained a selection of the many maps prepared for the Commission. The memorandum concluded with the following paragraph:

> The fundamental question that has to be decided is whether industry is to be forced or cajoled back into the old distributional pattern, or whether the industrial population is to be assisted to adjust itself to a new. With that problem, however, a Geographical Society has nothing to do. We are concerned rather to put the point that a series of national maps of the type which we have prepared and here put forward has a twofold value, in the first instance as presenting a clear picture of the geographical distribution of industry and the industrial population as it is today, and in the second instance as affording some guidance in respect of any future policy of planning.

Owing to the outbreak of war in 1939, the publication of the Commission's Report (Cmd. 6153) was delayed until 1940.

At their annual meeting at Cambridge in 1938, the British Association for the Advancement of Science appointed a committee of representatives from the several interested sections of the Association to draw up a scheme for such an atlas. Eva Taylor was appointed chairman of the committee. Later in the year the Royal Geographical Society nominated two members to serve on it. The outbreak of war in 1939 impeded the realization of the project. For more than a decade, Taylor was the leading protagonist for a national atlas of Britain. She never failed to remind any audience that statistical tables were of limited value unless also presented in map form. She wrote letters to *Nature* and accepted numerous engagements to lecture to organizations interested in planning. While always disclaiming expertise in the subject, she emphasized the importance of the geographical outlook and the desirability, indeed necessity, of a geographical basis for planning.

Early in 1941, the Government gave the Minister of Works (Lord Reith) responsibility for problems of reconstruction and shortly afterwards the words 'and Planning' were added to the title of his Ministry. Within his Ministry a small 'Reconstruction Group' was formed: this became the nucleus from which the Ministry of Town and Country Planning was formed in 1943. It had an advisory panel of which both Taylor and Stamp were members. In April, 1941 they submitted to Lord Reith an outline scheme for the initiation of a National Atlas in loose-leaf form and recommended that 'the first fascicule of maps should be those of immediate importance to planning and that the opinions of planners should guide in the selection'. Thus was initiated, under the guidance of E. C. Willatts at the Ministry of Works and Buildings, a series of maps, initially at the scale of 1:625,000 (and later the desk atlas of limited circulation). By 1944, the first instalment of ten maps had been compiled in the Ministry's Map Office and printed by the Ordnance Survey.

A decade after the atlas was first recommended, the Council of The Royal Society approved a proposal by the British National Committee for Geography for financial support from the Treasury. The latter's reply as recorded in a minute of a meeting of the British National Committee for 6 February 1950 indicates some of the difficulties placed in the way of those who advocated such an atlas. It was believed by HMSO that the price could not be less than eight guineas a copy – at which price it was suggested that fewer than the 20,000 copies apparently needed to 'break even' would be sold. The Treasury raised a further point against the proposal 'namely that an atlas containing so much industrial and economic material might be rather too informative to be published at the present time' (early 1950).

More attention was paid to the Royal Geographical Society's Memorandum than might have been expected because of the recognition by the drafting committee of the so-called 'axial belt' of high industrial population and manufacturing industry between London and Liverpool. Within a few years, several British geographers became alarmed at what they themselves believed was a 'doctrine' of an axial belt of industry in England. Although Taylor herself never regarded it as either a 'doctrine' or a 'theory', she asserted that it was a useful factual generalization of the order of Mackinder's division of Britain into 'Highland and Lowland zones'. The 'axial belt' or 'coffin' became a matter of considerable controversy and led to a number of heated debates. Among these was the discussion that followed the paper presented at the Royal Geographical Society some six years after the Memorandum had been laid before the Commission (Baker and Gilbert 1944). In their paper, J. N. L. Baker and E. W. Gilbert detailed many of those who had 'adopted the concept'. They also accused Professor Taylor of having 'adapted' Professor Fawcett's 'zone' of maximum concentration of population 'to fit her theory of an axial belt'. They also elaborated on the steps by which the 'theory' had become a 'doctrine'. Professor Taylor's actual reply to the authors was longer and more vitriolic than the brief printed reply (edited by the

Society's Secretary, A. R. Hinks[4]) would suggest. An indication of what she actually said (remembered by those in the audience still alive) can be gleaned from Baker's restrained reply to her outburst.

> I have not a great deal to say in reply because I do not think it is necessary to introduce a number of personalities into an academic discussion. We all know Professor Taylor and we make allowances. I am only sorry that she has gone before I had the opportunity of pointing out to her that she makes the best of both worlds: for part of the time within the axial belt, and for part of the time outside in the delightful country of rural England. So that she can have it whichever way she likes ...

Professor Taylor is on printed record as saying merely:

> As I have not seen the complete paper, and have only listened to it sitting in the dark, I cannot reply in detail. There is no 'doctrine' or 'theory' of an axial belt. It is a factual generalization of the order of Sir Halford Mackinder's division of Britain into a Highland and Lowland Zone which has proved useful. There is no suggestion in either case that such zones are unbroken. The slide showing new factory building between 1933 and 1937 and that showing the railway lines with four sets of tracks, with many that have been exhibited at the Society, confirm the existence of a belt running from south-east to north-west with a high degree of accessibility and therefore of attraction for industry.

The last recorded contributor to the discussion, A. E. Smailes, voiced aloud a significant change in the descriptive term for the axial belt or so-called coffin. He stated:

> I am particularly interested in the emphasis that has been laid in this paper upon the break in the centre of the so-called axial belt because my own studies have led me to modify the conception of a coffin-shaped area as put forward by Professor Taylor ... I prefer therefore to liken the shape of the main area of concentration of economic activity and population in Britain to an hour-glass, with its axis running through London and Manchester and its waist about Northampton.

The printed contributions of others to the discussion – S. H. Beaver, C. B. Fawcett, M. P. Fogarty (an economist at Nuffield College, Oxford), L. D. Stamp and E. C. Willatts – are interesting to read even after a lapse of nearly forty years. Indeed the text of the paper and the discussion give an insight into the 'applied geography' of the 1930s and 1940s and also provide an example of the 'cut and thrust' of academic discussion of the period. Academic geographers were very few in number – probably less than twenty per cent of the number in the 1980s – but many of them spoke with authority and were very active, bearing in mind their many commitments and their small number.

Eva Taylor retired from the chair of geography at Birkbeck College in 1944 at the then normal retirement age of 65. One could perhaps argue that she was one of the first British geographers to see the social relevance of at least one branch of geography. She had always advocated that, just as there was 'pure' and 'applied' mathematics, so there should be 'pure' and 'applied' geography. She also believed that geography should be a post-first degree subject; in other words that would-be students of geography should first be trained in another discipline.

During the Second World War, Eva Taylor devoted a considerable amount of her time to post-war planning. She was an active participant in the work of the Association for Planning and Regional Reconstruction which was guided by Jacqueline Tyrwhitt and the Earl of Verulam. In 1942, she opened a discussion on the geographical aspects of regional planning at the Royal Geographical Society (Taylor 1942a) and, in the same year, published a *Ground Plan of Britain* (Taylor 1942b). She also lectured on the geographical background of planning to a number of organizations including a summer school on Town and Country Planning organized by the Town Planning Institute in 1943 (Taylor 1943). In 1948, she gave evidence to the 'Schuster Committee on the Qualifications for Planners'. She persuaded the Town Planning Institute to include a paper in economic geography and applied geology, thus radically changing its professional examination. In 1950 she contributed five chapters to the *Town and Country Planning Textbook* edited by the Association of Planning and Regional Reconstruction and published by the Architectural Press (Taylor 1950).

Eva Taylor was not universally popular among her contemporaries. She combined a sparkling wit with an uncanny gift of putting her finger on the weakness in anyone else's argument and a tactlessness which endeared her to some and alienated others. Her reviews of the books of others were often vitriolic. Many were 'toned down' by the editors of the journals in which they were published but some were not.

During the inter-war years geography in Britain was largely the study of areal differentiation in man–land relationships. It was anchored to a physical basis. In spite of the pleadings of Unstead, geographers persisted in breaking-down areas according to the distribution of geological outcrops, static settlement patterns, etc., instead of analysing the regional association of these phenomena as distinctive landscape units of different magnitude. Both Unstead and Eva Taylor appreciated the value of field observation – both regarded the field as the geographer's laboratory. Theirs was field work in the W. M. Davis tradition. Neither forgot his demonstrations on the ground during the Oxford Summer School in 1904.

56   *Eila M. J. Campbell*

Both believed in the virtue of fieldwork on foot, with map in hand, and a keen eye on the landscape.

NOTES

[1] The chair of geography at Birkbeck College is listed by Fleure as the fifth chair to be established in Britain (Dickinson 1976).
[2] For a discussion in English of Passarge's ideas on the unit area and the hierarchy of such units, see Dickinson (1969).
[3] A complete bibliography of her works was published in 1968 in *The Transactions of the Institute of British Geographers*, 45, 181–6.
[4] Within a year of L. P. Kirwan's appointment to the post of secretary in 1945, his title was changed to that of Secretary and Director.

REFERENCES

J. N. L. Baker and E. W. Gilbert (1944), 'The doctrine of an axial belt of industry in England', *Geographical Journal*, 103, 49–72.
R. E. Dickinson (1969), *The Makers of Modern Geography*.
   (1976), *Regional concept: The Anglo-American leaders*. Excerpts of the Geographical Association's report on the classification of regions are printed on pp. 144–52.
Geographical Association (1937), 'Report on classification of regions of the world', *Geography*, 22, 253–82.
Richard Hartshorne (1939), 'The nature of geography', *Annals of the Association of American Geographers*, 29, 171–658.
A. J. Herbertson (1905), 'The major natural regions: an essay in systematic geography', *Geographical Journal*, 25, 300–12. Portions of this paper were reprinted under the title 'The natural regions of the world' in *Geographical Teacher* 3, 104–13.
   (1913a), 'Natural regions of the world', *Report of the British Association, 1913 Meeting in Birmingham*, 557–9.
   (1913b), 'The higher units: a geographical essay', *Scientia*, 14, 203–212. Reprinted, 1965, in *Geography*, 50, 332–342.
   (1914), 'Types and orders of natural regions', *Geographical Teacher*, 7, 160–3.
O. Maull (1936), 'Allgemeine vergleichende Länderkunde' in *Länderkundliche Forschung', Krebsfestschrift*, Stuttgart, 172–86.
A. P. Newton (ed. and contrib.) (1926), *Travel and Travellers of the Middle Ages*.
P. H. Schmidt (1937), *Philosophische Erdkunde: Die Gedankenwelt der Geographie und ihre nationalen Aufgaben*.
E. G. R. Taylor (1930), *Tudor Geography 1485–1583*.
   (1932a), Edition of *A briefe summe of geographie* by Roger Barlow. Hakluyt Society Second Series 69.

(1932b), Statement reported in the report of the meeting under the title 'What is Historical Geography?' in *Geography*, 17, 42.

(1934), *Late Tudor and Early Stuart Geography 1583–1650*.

(1935), *Edition of The Original Writings and Correspondence of the Two Richard Hakluyts*, 2 vols, Hakluyt Society Second Series, 76, 77.

(1936a), 'Leland's England', in H. C. Darby (ed.), *An Historical Geography of England before AD 1800*.

(1963b), 'Camden's England', in H. C. Darby (ed.), *An Historical Geography of England before AD 1800*.

(1938a), 'The geographical distribution of industry (being a paper to open discussion)', *Geographical Journal*, 92, 22–39.

(1938b), The 'Memorandum on the geographical factors relevant to the location of industry' (with maps and an appendix by Gordon Manley) in *Geographical Journal*, 92, 499–526.

(1939), 'The location of industry in Great Britain', *Geographical Journal*, 94, 334–5.

(1940), 'Plans for a national atlas (being a paper to open discussion)', *Geographical Journal*, 95, 90–8.

(1942a), 'Geographical aspects of regional planning (being a paper to open discussion)', *Geographical Journal*, 99, 61–3.

(1942b), *Ground Plan of Britain; fourteen maps for the use of planning authorities and groups* (with text).

(1943), 'The geographical background of planning', *Report of the Town and Country Planning Summer School 1943*, The Town Planning Institute.

(1947), 'Geography in war and peace', *Advancement of Science*, 51, 187–94. Reprinted in the *Geographical Review* (1948), 38, 132–41 and the *Scottish Geographical Magazine* (1947), 63, 97–108.

(1950), Contributions to Section 1, Geography in *Town and country planning textbook* edited by Association of Planning and Regional Reconstruction, London, The Architectural Press.

(1954), *The Mathematical Practitioners of Tudor and Stuart England*.

(1956), *The Haven-finding Art*.

(1966), *The Mathematical Practitioners of Hanoverian England*.

J. F. Unstead (1916), 'A synthetic method of determining geographical regions', *Geographical Journal*, 48, 230–49.

(1926), 'Geographical regions illustrated by reference to the Iberian Peninsula', *Scottish Geographical Magazine*, 42, 159–70.

(1932), 'The Lötschental: a regional study', *Geographical Journal*, 79, 298–311.

(1933), 'A system of regional geography', *Geography*, 18, 175–87.

(1935), *A Systematic Regional Geography. Volume I, The British Isles*.

J. F. Unstead and E. G. R. Taylor (1910), *General and Regional Geography for Students*.

5   The Oxford School of Geography

ROBERT W. STEEL*

The role of individual departments of geography, notably in the years immediately after the First World War, is outlined in several of the essays in this volume. The special position of Cambridge has been emphasized by more than one writer, and the importance of the University of London – with its close relationships with university colleges in a number of places, including Exeter, Hull, Leicester, Nottingham, Reading and South-ampton – will have been made obvious in other essays. The School of Geography in Oxford, with which the writer was associated from 1934 as an undergraduate and later as a member of staff until 1956, also made very significant contributions to the development of the subject in the inter-war period. The geographical tradition in Oxford is indeed as old in Oxford as in any other British university. The history of geography in Oxford has been described by, among other people, J. N. L. Baker (1963), E. W. Gilbert (1972) and D. I. Scargill (1976), and it is, therefore, unnecessary even to summarize it here. An appropriate starting point is the establishment in 1887 of the Readership in Geography held by H. J. (later Sir Halford) Mackinder. This was made possible by the genero-sity of the Royal Geographical Society which provided money for the appointment, largely as a result of the publication of the Scott Keltie Report on 'Geography in Education' in 1886 (Gilbert 1972; Scargill 1976). H. J. Mackinder was elected the first Reader. Although only twenty-six years old he had already made his mark, notably by his Oxford

* Robert Walter Steel, C.B.E. (b. 31 July 1915) graduated in the Honour School of Geography of the University of Oxford in 1937. He was appointed to the staff of the School of Geography in 1939. In 1957 he was elected to the John Rankin Chair of Geography in the University of Liverpool. From 1974 until his retirement in 1982 he was Principal of the University College of Swansea, and between 1979 and 1981 was Vice-Chancellor of the University of Wales. He became an Emeritus Professor of the University of Wales in 1982 and is an Honorary Fellow of Jesus College, Oxford.

University Extension courses on the 'new geography' and through his paper on 'the scope and methods of geography' that he read to the Royal Geographical Society in 1887 (Mackinder 1887). The then President of the Society, Sir Clements Markham, described him as 'a geographer of exceptional ability and great power of expression'; and many years later, after his death in 1947, J. N. L. Baker wrote of him (Baker 1947: 15) 'as the founder of the present School of Geography at Oxford, as a brilliant exponent of political geography, and as a master of the English language', to whom 'all his successors at Oxford are particularly in his debt' and 'whose inspiring work and stimulating ideas have done so much to advance the cause of British Geography'.

The School of Geography was founded twelve years later, in 1899, with Mackinder as its head. As his assistant, A. J. Herbertson, a zoologist trained in Edinburgh, was appointed. Later Herbertson was given the personal title of Professor of Geography for five years prior to his untimely death in 1915 (Gilbert 1972). But despite the support given to the University by the Royal Geographical Society there was considerable opposition to the establishment of a Chair of Geography from various sections of the University, and it was only in 1931 – three years after the comparable event in Cambridge – that the professorship came into being.

The School of Geography was originally housed in rooms on the upper floor of the Old Ashmolean Museum and then, on the opposite side of Broad Street, in Acland House. In 1921 the School moved into Holywell House in Mansfield Road, formerly the home of a Fellow of Balliol College. This is the building that still houses the Oxford geographers of the 1980s though there have been extensive additions, particularly in recent years. After the end of the First World War the School continued its policy of providing courses for the Diploma of Geography and arranging Summer Schools that were designed particularly for teachers, and of especial value for those whose initial training had been in a discipline other than geography. The organizers of these activities were the lecturers attached to the School of Geography, notably H. O. Beckit and James Cossar. Beckit, first appointed as a Reader in 1918, was Head of the School. He was a geomorphologist (though that word was never used in those days) and published a number of papers expounding concepts in physical geography. One of these, which became a classic essay on the evolution of the landforms of the Oxford region, appeared in the British Association handbook for 1926 (Beckit 1926). Cossar, first assistant to the Reader and later lecturer in geography, played a special role in relation to geography in education as exemplified in the Diploma courses. J. N. L. Baker joined the staff as assistant to the Reader in 1923.

Several geographers came regularly to assist in the Summer Schools, including E. G. R. Taylor, C. B. Fawcett and A. G. Ogilvie, all of whom had received the Diploma in pre-war years; many of those attending the Summer Schools became successful university or school teachers, and active members of bodies such as the Geographical Association and, after its foundation in 1933, of the Institute of British Geographers. They included G. H. J. Daysh, later associated so closely with the development of geography in Newcastle upon Tyne, and members of the Oxford staff of the thirties who had initially graduated in other subjects (chiefly Modern History) – C. F. W. R. Gullick and E. W. Gilbert in addition to Baker.

The establishment of the Chair and of an Honour School of Geography – so vital to the proper recognition of geography as an academic discipline by the University – faced many difficulties in an essentially conservative and traditional university. Thus in 1913 a proposal to establish an Honour School of Geography (as part of the Honour School of Natural Science) was unsuccessful, and the same fate met another proposal put forward in 1918 (Firth 1918; Baker 1963: 127). Progress was also slow during the twenties and it would never have come about but for the keen support of J. L. (later Sir John) Myres, the Camden Professor of Ancient History. In earlier years, when he was Professor of Ancient History in the University of Liverpool, he had done much to help to create the atmosphere in which Liverpool embarked upon the establishment of an Honour School of Geography (the first in any British university) and the founding of the John Rankin Chair of Geography, to which P. M. Roxby was appointed in 1917 (Steel 1984: 2–3). Back in Oxford Myres campaigned resolutely, with others, mainly classicists and historians, for the establishment of a similar chair. His own *Dawn of History* (1911) was an outstanding example of geographical work done by an ancient historian, and he had greatly extended his knowledge of the geography of the Aegean through his activities in the First World War as a naval commander (many years later he compiled, largely without assistance, the Naval Intelligence Geographical Handbook on the Dodecanese Islands, published in 1943). The Diploma Course in Geography had by now served its purpose but the numbers taking it were beginning to decline since several universities were now offering honours degrees in the subject. Eventually the case for the establishment of a chair of geography and of an associated Honour School was accepted during 1929 and 1930 by the University. The Royal Geographical Society lent its powerful support, this in pursuance of its policy more than four decades earlier when it had established the Readership, and the first examination for the new School was held in 1933.

The Chair was founded in 1931 and filled a few months later with the election of Major Kenneth Mason, M.C., who had joined the Royal Engineers from the Royal Military Academy, Woolwich, in 1906. He had long professional experience as a surveyor with the Survey of India both before and after the First World War, and for his scientific work, particularly in the Karakoram, he was awarded the Royal Geographical Society's Cuthbert Peek Grant in 1926 followed by the Founder's Medal in 1927. His study of geomorphological processes in the Shaksgam valley was published in the *Geographical Journal* in 1926 (Mason 1926).

A senior colleague was Nora E. MacMunn, who had taken the Diploma in 1904 and been appointed a demonstrator in the School of Geography in 1906, and who remained a member of the staff until her retirement in 1935. As a true disciple of Herbertson she delivered an annual course of lectures, extending over all three terms, on 'Natural Regions'. Baker had, after a gap for war service, completed the Honour School of Modern History in 1920 and then took the Diploma in Geography course in 1921 prior to being awarded the B.Litt. research degree for a thesis on 'Geographical aspects of the Peninsular War'. He spent a year in the University of London at Bedford College before returning to Oxford in 1923. His interests were always markedly on the historical side of geography and his standard textbook, *A History of Geographical Discovery and Exploration*, first appeared in 1931. His pupil, and later colleague, C. F. W. R. Gullick, was another history graduate who then took the Diploma course and undertook research which led to the degree of B.Litt. for a thesis on the geographical development of West Cornwall. Another historian was E. W. Gilbert, who became Oxford's Professor of Geography from 1953 to 1967. He was awarded the Diploma in 1924 and did research in historical geography. His B.Litt. thesis (1928) was published as a book in 1933 with the title *The Exploration of Western America, 1800–1850: an historical geography*. Like Baker, he taught first at Bedford College, University of London, and then worked, with A. Austin Miller, at University College, Reading (after 1929 the University of Reading), returning to Oxford as Research Lecturer in Human Geography in 1936.

There is no doubt that inter-war undergraduates derived considerable benefit, in Oxford as elsewhere, from the initial training that the geographers who taught them had had in other disciplines. In Oxford most of the full-time teachers had been trained as historians, but we profited equally from the fact that several of our teachers (especially in the field of physical geography) were expert in other fields. These included W. G. Kendrew (a classicist with a special interest in climatology and widely read by geographers for many years because of his books, *Climate* (1930) (re-titled from 1949 as *Climatology*) and *The Climates of the Continents* (1922)); K. S. Sandford (geology) whose courses on landforms were always given in the School; A. G. Tansley (Professor of Botany), who was one of the pioneers of the concept of ecology and the author of *Types of British Vegetation* which first appeared in 1911; Colonel M. O'C. Tandy, a retired member of the Survey of India, who conducted practical classes in the University Parks and appeared to be quite oblivious to the counter attraction of first-class cricket during the Summer Term; and L. H. D. Buxton, a physical anthropologist, who was a regular lecturer in the School of Geography up to the time of his sudden death in 1939 and whose presence stressed the significance of the relationship between geography and anthropology as suggested by the establishment of the Faculty of Anthropology and Geography. Their students were encouraged to read widely, few concessions being made to the lack of background in these subjects they might have as undergraduates who had studied geography in the sixth form; and in fact quite a number reading for the Honour School of Geography had not even done that and for them geography was a wholly new subject.

When the Honour School of Geography began in 1931 there was already a nucleus of trained geographers in Oxford and some useful links with schools (though perhaps more with grammar than public schools). The first candidates were examined in June 1933; there were only two, both of whom were awarded Thirds. In 1934 there were eleven candidates (four of them women), and Dorothy M. Doveton was placed in the First Class and was awarded a Drapers' Company research scholarship. This enabled her to visit Swaziland and later to produce one of the first monographs published by the Institute of British Geographers, *The Human Geography of Swaziland* (1937). In 1935 fourteen candidates (five women) sat for the examination, and G. E. Holderness, who subsequently became Bishop of Burnley, was awarded a First. Thereafter numbers increased steadily from twenty in 1936 to thirty-one and thirty-five in the two following years, and the inter-war record of thirty-nine – from twenty-one different colleges – was reached in 1939 when there were nine women candidates and four Firsts were awarded.

What books were Oxford undergraduates expected to read during the thirties? Freeman (1980: 185) has written: 'It will seem strange to students of the 1970s that fifty years earlier there was so little to read on many aspects of geography that one welcomed anything new with joy. A new article by P. M. Roxby on China, the 1928 volume of essays on Great Britain edited by A. G. Ogilvie, the "Corridors of Time" volumes by

H. J. E. Peake and H. J. Fleure, and other books of that epoch were eagerly read, as were the small number of American texts, not least Isaiah Bowman's fascinating *New World*. French texts were also eagerly read, by many students in the original language.' But such was the paucity of geographical books at this time that I think that I had probably read most, if not all, of the books specifically written by British geographers by the time I graduated in 1937. This was in addition to classical American texts such as Bowman's *The New World: problems in political geography* (first published in 1921), to which Freeman has referred, W. M. Davis's *Geographical Essays* (1909) (the essays on education as well as the better-known physiographical essays, making a volume of 777 pages in all), D. W. Johnson's *Shore Processes and Shoreline Development* (1919), and some of Ellsworth Huntington's stimulating if provocative works, such as *The Pulse of Asia* (1907) and *Civilization and Climate* (1915). We were also encouraged – as Freeman had been a few years earlier – to read some of the best-known works in French (a few of which were available in translation), by authors such as Emmanuel de Martonne, Vidal de la Blache and Jean Brunhes together with the volumes of the famous, though now, I suspect, all-but-forgotten volumes of the *Géographie Universelle* series. Indeed all taking the Honour School of Geography had to study, as a region, either Central Europe or the Mediterranean Lands, based on a set-book in this series. Thus we worked through selected parts of de Martonne's *Europe Centrale* (1930) or the volume on *Méditerranée* by J. Sion and Y. Chataigneau (1934). To ensure that we studied these works with care, we were examined with passages for translation and for interpretation by the setting of 'gobbets' (a peculiarly Oxford term for extracts from set texts which needed both translation and comment). The reading of these volumes helped us to appreciate the basis for our teachers' high regard for the French School of Geography and in particular for the French concept of regional geography which stood in high favour in those days, and not only in Oxford.

The comparative lack of books written by geographers had a further effect upon us, and our tutors, for we were strongly advised to read outside the subject to an extent that is seldom true today. Students in the eighties read widely – some of them at least – but so often they are recommended not to read what geologists, economists, historians and social anthropologists have written about their disciplines but what geographers, their own teachers and others, have interpreted as the intentions of these other scholars. We were encouraged, for example, to read books such as Mahon's *Influence of Sea-Power upon History* and M. Bloch's work on the organization of agriculture in France, to mention but two authors whom I recall studying with considerable profit; and since there were relatively few 'geographical' periodicals at the time (no *Transactions* of the Institute of British Geographers, no *Area*, no *Progress in Physical Geography*, no *Progress in Human Geography*, no *Journal of Historical Geography* and with a *Geographical Journal* that was much more directed to exploration and discovery than it is today), we were expected to read papers regularly in journals such as the *Quarterly Journal of the Geological Society*, the *Journal of Ecology*, *Economica*, *International Affairs*, *Foreign Affairs* and *Archaeologica* – again to mention only a selection of those that were taken by the library of the School of Geography which, in my view, remains to this day one of the best-stocked libraries for geographers in Britain.

It would be tedious to list all the books written by British geographers that were available in those days even though they were few in number compared with the output of books today. There was the Methuen series of (predominantly) regional texts – some of them admittedly very dull and turgid, and packed with detailed factual information. Yet I believe that those of us who read Hilda Ormsby's *France: a regional and economic geography* (1931) 'knew' France and had an understanding of that country and its people (and their very considerable inter-war problems) that I suspect few geography students have today; and Walter Fitzgerald's *Africa: a social, economic and political geography of its major regions* (1934) played an important part in turning my thoughts as a research student to that continent even though only five of its 462 pages were devoted to Sierra Leone, the country to which I went in 1938 to do my fieldwork in preparation of a thesis on 'The human geography of Sierra Leone'. I know, too, that L. Dudley Stamp's *Asia: a regional and economic geography* (1929 and subsequent editions) – dedicated, it is worth remembering, to Mrs Elsa Stamp, herself a geographer, 'in memory of bullock-cart days and Irrawaddy nights' – had a profound influence on many of us, Indian and non-Indian alike, who have concerned ourselves with that country's overwhelming problems in subsequent years.

Also important in the Methuen series, though not a regional volume, was A. Austin Miller's *Climatology* though students also had available W. G. Kendrew's two volumes to which reference has already been made. Two books published while I was still at school in the sixth form had a very considerable influence upon my career – *The British Isles: a geographic and economic survey* by L. Dudley Stamp and S. H. Beaver and C. B. Fawcett's *A Political Geography of the British Empire* (1933). Others besides Beaver who have also contributed to this volume and who published important books during this inter-war period are J. A.

Steers and H. C. Darby. The former's *An Introduction to the Study of Map Projections* (1926) was valued by undergraduates at a time when all geography students were expected to have at least a working knowledge of map projections. Darby's editorship of *An Historical Geography of England before AD 1800* (1936) opened a whole new world of enquiry to many of us, with all but one of the fourteen studies written by geographers. (The exception was the essay on Scandinavian settlement by Eilert Ekwall, Professor of the English Language in the University of Lund (see p. 125).) Another important book of the late thirties was *The Physical Basis of Geography: an outline of geomorphology* by S. W. Wooldridge and R. S. Morgan (1937) which was published only a few weeks before I took my finals. This was for many of us our first introduction to the now long-accepted term geomorphology (physiography was still the word in most common usage at the time). It also underlined the importance given in the University of London syllabus of this period, and indeed for many years afterwards, to what was described as 'the physical basis of geography'.

Individual reading rather than attendance at lectures was perhaps given special emphasis in Oxford as the essential background for the production of the weekly essay for one's tutor, but most of the activities of the Oxford School of Geography would have been repeated in many other departments. We all did a course on geological mapping in the Department of Geology, and were expected to master at least the rudiments of surveying (which was examined in the Final Honour School of Geography). Map projections, however, were not regarded as important as in the Cambridge Tripos syllabus. There was no formal teaching in cartography – indeed freehand mapping was almost encouraged – and it was not until after the war that texts began to appear such as *Maps and Diagrams* by F. J. Monkhouse and H. R. Wilkinson (1951). This book was in fact the published version of the map classes that they gave in Liverpool with the very strong encouragement of the then Head of the Department of Geography, H. C. Darby.

Students were encouraged to attend lectures delivered elsewhere than in the School of Geography or by teachers from other disciplines who came to the School to give specific courses. I recall two series in particular: One was given by Dr Marett, a son of R. R. Marett, who had been Rector of Exeter College and was a close associate of the geographer H. J. Fleure, an introductory course on population – always delivered on Saturday mornings, no doubt to test the devotion of his audience which was usually small but invariably included some nuns who, some of us felt, were probably rather embarrassed by the nature of some of the topics that he chose to discuss. Another lecture course that had a very significant influence on my later career and my interest in tropical Africa, was a course of lectures on 'the economic geography of British West Africa'. These were given by T. M. Knox, then a philosophy don and a Fellow of Jesus, my own College, but later Principal and Vice-Chancellor of St Andrews University, where he was knighted as Sir Malcolm Knox. He drew very interestingly, and with a remarkable degree of detail, on his experiences as Secretary to Lord Leverhulme between 1920 and 1925 when he spent considerable periods of time in West Africa, the centre of the most important commercial activities of The United Africa Company, a major Unilever subsidiary.

There were also occasional series of university lectures for anyone interested. These were usually delivered in the Examination Schools between tea and dinner and were arranged by particular professors. Sometimes there were several distinguished lecturers (a different one each week) or a short series might be given by a well known scholar from another university. Thus an important series on the regional problems of the British Isles in the difficult years of the thirties had lecturers from Oxford such as G. D. H. Cole and M. Fogarty, or authorities from elsewhere. My particular recollection is of Professor J. F. (later Sir Frederick) Rees talking about the problems of South Wales. Speakers such as Sir Gilbert Murray and Sir Alfred Zimmern spoke on important international topics, including the problems and prospects of the League of Nations and the dangers of the armaments race. Such lectures provided undergraduates and their teachers with an opportunity of hearing lecturers with great reputations. We in Oxford were perhaps especially privileged though undergraduates in Cambridge and in London, where the tradition of inter-collegiate lectures has always been so established, would have enjoyed similar advantages.

With the economic recession of the early thirties and the worsening of the international situation during the years before the outbreak of the Second World War in 1939 – with the Japanese attacks on China, the Italian invasion of Ethiopia (then generally known as Abyssinia) and Nazi aggression in various parts of Europe, and the stepping-up of the armament programmes in so many countries – the geographers of the time recognized the need to stress the relevance of geography in both national affairs and international relations. Liverpool's concern about the Chinese situation in Manchuria, for example, was especially marked and was a direct reflection on the number of Chinese students who came to Merseyside to work under the guidance of P. M. Roxby; while in London C. B. Fawcett underlined the importance of many social, economic and political problems in his comprehensive survey of the political geography of the British Empire (1934).

The involvement of geographers in social and economic issues within the British Isles is particularly well illustrated by the work of L. D. Stamp and others in the London School of Economics (see the essays in this volume by S. H. Beaver and E. C. Willatts), while the Royal Geographical Society interested itself in the preparation of background material for the Barlow Commission on the distribution of the industrial population and, subsequently, for a National Atlas (see the essays by Willatts and E. M. J. Campbell in this volume, and also Taylor, 1940). From Oxford there was a notable contribution by E. W. Gilbert in the first volume of *A Survey of the Social Services in the Oxford District* (1938). The Oxford School of Geography indeed laid considerable emphasis on the relevance of geographical training to the understanding of current affairs which Mason had stressed in his inaugural lecture on 'The geography of current affairs'. This represented not only his thinking (for he subsequently lectured each year on current affairs) but also that of his most senior colleague, Baker. Perhaps it was also a reflection of the continued influence of H. J. Mackinder, the first Reader in Oxford, for during the inter-war years the Oxford School of Geography was probably one of the few departments where students were encouraged to read his *Democratic Ideals and Reality* (published in 1919) as well as his earlier papers, such as 'The geographical pivot of history' (*Geographical Journal*, 23 (1904), 421–44). These were widely read, in the USA as well as in Britain, during and after the war with the newly found interest in geopolitics (and the use made of it by Nazi Germany). Oxford's concern with the relationships between geography and politics is particularly illustrated in Baker's paper 'Geography and politics: the geographical doctrine of balance' (Baker 1947). This appeared in 1947 at the end of his period as President of the Institute of British Geographers, though it was not a presidential address as such (Steel 1984: 85). Earlier, however, he had indicated his interest in current affairs by his editorship of one of the first (no. 22) of the series of 'Oxford Pamphlets on World Affairs', *An Atlas of the War* (1940), which appeared, with the help of three of his colleagues (E. W. Gilbert, C. F. W. R. Gullick and R. W. Steel), in the early months of the Second World War. Even before then there is evidence of his desire to collaborate with R. O. Buchanan in the preparation of an atlas of economic and political geography under the auspices of the Institute of British Geographers, though in fact nothing came of their plans for such a publication (Steel 1984: 20).

Baker's key role in the development of geography in Oxford was also shown by his skilful operations behind the scenes, in the University and more particularly in his own college, Jesus. Very shortly after the establishment of the Honour School of Geography, Jesus College began to offer an award for geography though initially it was for geography in competition with modern languages and English. Closed Scholarships (Meyricke) were offered to those with appropriate Welsh qualifications (Jesus having then, as now, special links with Wales) though the first award (and several subsequent ones) were of open exhibitions to those without the necessary Welsh background. It was only considerably later that other colleges offered any specific awards for the subject, and right up to the outbreak of war geography still ranked in the Oxford hierarchy of subjects near the end of a queue that began with classics, theology and mathematics. There were no Fellows in Geography in any College apart from Mason, who as the Professor of Geography held a Professorial Fellowship at Hertford College. Baker was a very active member of the Senior Common Room at Jesus where his official title was 'Lecturer in Geography'; it was only after he became Bursar of the College in 1939 that he was elected into a Fellowship. The large increase in the number of geography Fellows in different Colleges came after the war, and then only gradually. Oxford being a college-dominated university, the members of the staff of the School of Geography were thus able to make only a relatively small contribution to the life of the university as a whole. But their impact on the geographical world was considerable. Baker, for example, was a most loyal and active member of Section E (Geography) of the British Association for the Advancement of Science and was Recorder of the Section from 1936 to 1949. He was also a founder member of the Institute of British Geographers and but for him and R. O. Buchanan and one or two other geographers, the Institute might never have revived from the inevitably dormant period of the war years (Steel 1984: 23). Mason was a Vice-President of the Royal Geographical Society and was very active in its work, and for many years (1928–45) he was editor of the *Himalayan Journal*. He took great pride in the fact that members of his family had been closely associated with the Drapers' Company, one of the City of London livery companies, since the fifteenth century and he himself was elected Master of the Company in 1949. From the standpoint of geography his connection with the Drapers' Company was very significant for he was instrumental in persuading the Company to give money that enabled three graduates of the School of Geography to do research in different parts of the Commonwealth – D. M. Doveton (subsequently Mrs Dicey), to whose work in Swaziland reference has already been made; A. F. Martin who spent some months in Newfoundland in

1937–8; and R. W. Steel who carried out fieldwork in Sierra Leone in 1938.

At an earlier stage the Oxford School had been responsible for the production of many textbooks, mainly written by A. J. Herbertson and his wife, and these continued to be widely used at least during the first half of the inter-War period. *Europe: a regional geography* by N. E. MacMunn and G. Costar, first published in 1923, also remained a standard school text for many years, and C. C. Carter, who came to the School after his retirement from many years of teaching at Marlborough College, published his *Landforms and Life* (1938) after his move to Oxford, thereby extending still more the great influence on the development of geography that he had established not only through his reputation at Marlborough and among public schoolmasters concerned with geography but also by his other published works, including *A Geographical Grammar* (1929) and (with H. C. Brentnall) *The Marlborough Country* (1932). Apart from the books by Baker, Gilbert and Kendrew noted above (pp. 61–2), the Oxford School produced little in the way of research publications but that was common to most of the departments of geography of the time, with their small and often over-worked members of staff, and with the outlets for publication very limited. Prospects of academic geographers obtaining space in the *Geographical Journal* were in those days meagre, while *Geography*, the publication of the Geographical Association, very understandably concentrated on work that was of direct relevance to those teaching in schools. Even with the foundation of the Institute of British Geographers in 1933, with better facilities for publishing as one of its main aims, the Institute's policy until after the war was to publish monographs rather than papers (Steel 1984: 56). Worthy of note as an Oxford publication, however, is the quite slim volume prepared by C. F. W. R. Gullick and published as *The Oxford District* in 1939 as the first in a series, 'A pictorial survey of England and Wales', issued by George Philip and Son Limited. This, like others that appeared subsequently in the series,* was designed to make available to a wider public and to visitors to the areas covered, the outcome of the fieldwork undertaken by members of the departments of geography in their own regions into which they regularly took their students.

Fieldwork played a very significant part in the curriculum of geography in the inter-war period, not only in Oxford but in most other departments, reflecting very much the then current views that geography was supremely

* Only three were published, probably as a consequence of the outbreak of war shortly after Gullick's booklet appeared. The others of the series were concerned with Lancastria and with the Midlands.

concerned with first-hand experience and observation. Indeed quite apart from undergraduate involvement in such activities, field excursions formed an integral part of the university-organized Summer Schools, while the Le Play Society ran a series of courses in different countries, usually in some of the lesser known areas of Europe; and a number of geographers participated, including two of the contributors to this volume, S. H. Beaver and K. C. Edwards. Indeed the geographical content of these courses was probably greater between the wars than in the post-war period when the sociological interests of the founder, François Le Play, became more dominant, partly because geographers were becoming increasingly involved in other field activities such as those organized by the Geographical Association and by the Geographical Field Group. The latter was based at the University of Nottingham and many Le Play Society enthusiasts gravitated towards it (Edwards' involvement in this development is referred to on p. 97). The British Association for the Advancement of Science, and especially Section E (Geography), had always included fieldwork as an integral part of its programme, and in Section E the afternoon of the first full day was traditionally devoted to the geographical study of the city or town where the meeting was being held, while on the Saturday and the Sunday the local organizers were able to arrange longer full-day excursions under the leadership of some of the local geographers. Similarly the Institute of British Geographers right from its beginning in 1933 thought in terms of a serious, professionally organized programme of fieldwork, initially in very close association with the British Association summer programme, but increasingly under its own auspices (Steel 1984: 101–2). This tradition continued for some years after the end of the Second World War – though with increasing difficulty – and the fieldwork component in Institute of British Geographers' activities today is generally very slight indeed.

There were no minibuses or landrovers in departments in those days – and not many private cars. Often field excursions were undertaken on foot, perhaps with a starting point at a conveniently placed railway station. In Oxford – and no doubt elsewhere – bicycles were commonly used, even for some years after the end of the war. Cars were made available for more distant forays into the Cotswolds to see, among other things, examples of settlements with superb buildings – churches, farm buildings and houses – that expressed the medieval wealth of the area. In the opposite direction field classes went to the Berkshire Downs, especially to the White Horse at Uffington, and to the Chilterns, with a special eye for both the evolution of the Goring Gap and the present-day significance of the break through the hills used by modern communications between Oxford and Reading. Field excursions were also organized on the Continent, mostly in France or in Germany and lasting about ten days. They cost ridiculously little, or so it would seem today, usually about £15 for travel and accommodation. They were arranged by J. N. L. Baker, assisted by his colleagues, in every year from 1934 until the outbreak of war in 1939. It is perhaps noteworthy that the travel agency who made the arrangements, Bells Travel Service, was owned by and managed by a businessman, F. T. Holbrow, whose interest was so aroused by all that he did for the School of Geography that he became Chairman of the Oxford Branch of the Geographical Association and retained that post (which he loved and filled with distinction) for many years.

There was considerable emphasis, in Oxford as elsewhere, on the value of original work and to this end most departments asked for the submission of a dissertation undertaken by the student during vacations. Oxford has always been interested in the concept of H. R. Mill for an official regional survey of Britain based on the one-inch map sheets published by the Ordnance Survey (Mill 1900), and all Diploma students had to prepare a geographical account of one of these sheets. The best-known of these was that of the Andover sheet by O. G. S. Crawford since this was published in 1922 and became a classic for its exposition of the relationships between geography and archaeology (Crawford, 1922).* The boundaries of map sheets were arbitrary and before the Honour School of Geography was instituted (in 1931) the requirements had been modified to 'a geographical description of an area not exceeding 150 squares miles' with a word limit of between 10,000 and 15,000. The examiners looked for what would now seem a very conventional account of, first, the physical geography and then the human geography of the area chosen. In practice during the thirties we spent an excessively long time in determining precisely which parish should be included and which not (this for statistical purposes), and we were discouraged from following up in any detail specialist aspects of the area under consideration; while

---

* O. G. S. Crawford was awarded the Diploma in Geography in 1910. He became a leading archaeologist and was the first Archaeological Officer of the Ordnance Survey. He provides an interesting commentary on the standing of geography in pre-1914 Oxford in his autobiography, *Said and Done*, (1953: 44), when he describes the effect on his tutor of his decision to forsake the classics and the reading of 'Greats' to do the Diploma in Geography: 'It was like a son telling his father he had decided to marry a barmaid ... Going from Greats to Geography was like leaving the parlour for the basement; one lost caste but one did see life. Geography was then a new subject, struggling to gain recognition. It was inadequately housed in a couple of overcrowded rooms in the Old Ashmolean Building. I immediately felt at home in the new environment of maps and things of this world, so refreshingly different from the musty speculations about unreal problems that had hitherto been my fare.'

because of the difficulties of urban study, as they were seen at the time, we were dissuaded from including in our chosen area anything larger or more complex than a medium-sized market town. Thus my area, the Kennet valley below Hungerford, included Newbury but the area had to stop at the suburban edge of the County Borough of Reading! These regional studies, for all their shortcomings (as seen with hindsight), were both valuable and enjoyable exercises. The conscientious geographer spent many weeks in the selected area, acquiring considerable local knowledge usually through extensive travel by bicycle. The less motivated students, however, could produce a geographical description that 'passed muster' on the basis of a very hurried reconnaissance of the area, a quick survey of such literature as was available, and (one sometimes suspected) a judicious perusal of theses that had been prepared by earlier generations of students (this was especially true of 'popular' areas such as sections of the Weald or the Chilterns, the Isle of Purbeck, and parts of the Pennines or the Lake District).

This chapter, as with other essays in the volume, contains many personal reminiscences and reflections, with the emphasis heavily on the academic side of the Oxford School of Geography and on the undergraduates' involvement in it. No attempt has been made, for example, to discuss, except incidentally, the provision (and in general the complete absence) of the ancillary resources – secretarial, cartographic and technical services and laboratory and other equipment – that can be taken for granted in departments of geography today (even if they are still far from adequate in many universities and polytechnics). Nor is there any assessment of the influence in later years of the Oxford School of Geography in terms, for instance, of the appointment of its graduates as teachers of the subject in universities in Britain and overseas. Such an analysis does not form part of this volume which is concerned with the state of the subject, and of its role in universities, in the years between 1918 and 1945. Other writers will make their judgements on the roles of Oxford, and other departments, as they look at the progress of geography since the end of the Second World War, and for Oxford an especially appropriate time is the present as the School of Geography prepares to celebrate the centenary of H. J. Mackinder's appointment as Oxford's first Reader of Geography. Two observations may, none the less, be appropriate. First, it may be noted, that in the year 1920, according to D. I. Scargill, 'no fewer than 12 out of 35 recognized teachers of geography in the universities or university colleges in Great Britain were former students of the Oxford School, and six such geography departments were wholly so staffed' (Scargill 1976: 459). Secondly, the Oxford School may take some credit for the international reputation achieved in post-war years by one of its students from overseas. Chauncy D. Harris was a Rhodes Scholar at Lincoln College, Oxford, and read the Honour School of Geography between 1934 and 1936, when he graduated in the Second Class. He then went to the London School of Economics to do research and subsequently taught at the University of Chicago for many years. He was Secretary-General and Treasurer of the International Geographical Union (one of geography's most influential and prestigious positions) from 1968 to 1976.

Much of what has been written about Oxford, which is particularly well-known to the writer, could equally apply to many, perhaps most, of the university departments of geography in Britain during the inter-war period. This essay is offered, like others in the volume, for the consideration of a younger generation of geographers who have been used to rather different conditions and circumstances in the institutions with which they are familiar, and to a much more sophisticated view of the subject as an academic discipline than the geography we taught, or were taught, between the wars. Bearing in mind the small size of the departments of those days and the heavy pressure on the time and energies of many of the staff, and taking into account the general lack of support given to them in both human and physical resources, and even the suspicion in which geography was often held by those in high places, those active in the subject today may possibly feel with the writer and his colleagues that very commendable efforts were made by the few geographers of the inter-war years, not only in Oxford but in many other departments, certainly in all those departments represented by the authorship of this volume. Perhaps we should all remember what Arnold Bennett has called 'the fun and reward of Geography' for this was in the forefront of the thinking of geographers between the wars. Moreover as we reflect on the achievement of that period in laying solid foundations for the later advance of the subject, we may with advantage remind ourselves of T. W. Freeman's cautionary words in *The Geographer's Craft* (1967: 198), 'to condemn or ignore the workers of the past may be to thwart the workers of the future'.

REFERENCES

J. N. L. Baker (1947), 'Geography and politics: the geographical doctrine of balance', *Institute of British Geographers Transactions*, 13, 1–15.
  (1963), 'The history of geography in Oxford', in *The History of Geography: papers by J. N. L. Baker*, 119–29.

H. O. Beckit (1926), 'Physiography of the Oxford region', in J. J. Walker (ed.), *The Natural History of the Oxford District*.

O. G. S. Crawford (1922), *The Andover District* – an account of sheet 283 of the One-Inch Ordnance Map (small sheet series), 100 pp.

(1953), *Said and Done: the autobiography of an archaeologist*.

C. B. Fawcett (1934), *A Political Geography of the British Empire*.

C. H. Firth (1918), *The Oxford School of Geography*.

T. W. Freeman (1967), *The Geographer's Craft*.

E. W. Gilbert (1938), 'Geography' in A. F. G. Bourdillon (ed.), *A Survey of the Social Services of the Oxford district*, vol. 1, 1–24.

(1972), *British Pioneers in Geography*, especially chapter 1, 'Richard Hakluyt and his Oxford predecessors', 30–43 and chapter 10, 'Andrew John Herbertson (1865–1915)', 180–210.

E. W. Gilbert and W. H. Parker (1969), 'Mackinder's "Democratic ideals and reality" after fifty years', *Geographical Journal*, 135, 228–31.

A. J. Herbertson (1905), 'The major natural regions', *Geographical Journal*, 25, 300–12.

J. S. Keltie (1886), Report to the Council of the Royal Geographical Society, *Report on the Proceedings of the Society in Reference to the Improvement of Geographical Education*, 1–156.

W. G. Kendrew (1922), *The Climate of the Continents*.

(1930), *Climate: A treatise on the principles of weather and climate*. The third edition (1949) was retitled *Climatology: treated mainly in relation to distribution in time and place*.

K. Mason (1927), 'The Shaksgam Valley and Aghil Range', *Geographical Journal*, 69, 287–332.

H. J. Mackinder (1887), 'On the scope and methods of geography', *Proceedings of the Royal Geographical Society*, N.S. 9, 141–60.

(1904), 'The geographical pivot of history', *Geographical Journal*, 23, 421–37 (1919), *Democratic Ideals and Reality*.

N. E. Macmunn and G. Costar (1923), *Europe: a regional geography*.

H. R. Mill (1896), 'Proposed geographical description of the British Isles', *Geographical Journal*, 7, 345–65

(1900), 'A fragment of the geography of England: South-west Sussex', *Geographical Journal*, 15, 205–27, 353–78; 16, 246

J. L. Myres (1911), *The Dawn of History*.

D. I. Scargill (1976), 'The R. G. S. and the foundations of geography at Oxford', *Geographical Journal*, 142, 438–61. See also D. R. Stoddart.

R. W. Steel (1984), *The Institute of British Geographers: the first fifty years*.

R. W. Steel and R. Lawton (eds) (1967), 'Geography at the University of Liverpool', in R. W. Steel, *Liverpool Essays in Geography: a jubilee collection*, 1–23.

D. R. Stoddart (1975), 'The R. G. S. and the foundations of geography at Cambridge', *Geographical Journal*, 141, 216–39.

E. G. R. Taylor (1938), 'Geographical distribution of industry', *Geographical Journal*, 102, 22–39.

(1940), 'Plans for a national atlas', *Geographical Journal*, 105, 96–108.

6 Geography in the Joint School (London School of Economics and King's College)

S. H. BEAVER*

The first students for the new Honours Degree in Geography in the Faculty of Arts in the University of London started their courses in October 1918, one month before the end of the First World War. They graduated in 1921, by which time similar courses could also, with a different background of subjects at 'intermediate' level, lead to the degree of B.Sc. The teaching of geography in the University goes back much further, for H. J. Mackinder, then Reader in Geography at Oxford, was amongst the lecturers listed in the first prospectus of the London School of Economics in 1895 (he subsequently became Reader in 1908 and Professor in 1923), and L. W. Lyde's Chair of Economic Geography was established at University College in 1902.

By 1906 L.S.E. had established a Certificate in Geography, primarily for school-teachers; this was superseded in 1910 by the University's Academic Diploma in Geography, which was of full honours standard.

The new degrees called for a wider basis of instruction, particularly in physical, mathematical and historical geography, than was available at L.S.E., while at King's College facilities existed in physical geography (taught by the Professor of Geology), mathematical geography (taught in the Civil Engineering Department), and historical geography (there was a Professor of Imperial History), but none at all in regional or economic geography. It was perhaps natural, therefore, especially in view of the proximity of the two Colleges (off Aldwych and on the south side of the Strand) that thoughts should turn towards a pooling of resources,

* Stanley Henry Beaver (b. 11 August 1907) graduated in geography and geology at University College London in 1928. In 1929 he moved to the London School of Economics where he was successively Assistant Lecturer, Lecturer and from 1946 Sir Ernest Cassel Reader in Economic Geography. In 1950 he became the Foundation Professor of Geography in the University of Keele, a post he held until his retirement in 1974 when he was made an Emeritus Professor. He died on 10 November 1984 in Eccleshall, Staffordshire.

and negotiations between the Heads of the two Colleges, Sir William (later Lord) Beveridge and Dr (later Sir) Ernest Barker, aided and abetted by Professor W. T. Gordon (Geology) and Professor A. P. Newton (Imperial History) of King's College, and Sir Halford Mackinder and Major Ll. Rodwell Jones of L.S.E., led in 1922 to the foundation of the Joint School of Geography. Henceforth, economic and regional aspects of the subject would be taught at L.S.E. by Mackinder, Rodwell Jones, his elder sister, Hilda Ormsby (who had been Mackinder's assistant since 1912), the Professor of Commerce, A. J. Sargent (economic geography), and L. G. Robinson (historical geography), with P. W. Bryan as Assistant for the practical classes (succeeded in 1923 by Winifride Hunt), while the appropriate 'background' subjects on the physical, biological and mathematical sides were taught in the relevant departments at King's, with the Geology Department providing a 'home' for the geography students. Moreover, it must not be forgotten that geography could be taken as one of three subjects for the B.A. General and B.Sc. General degrees, and that it was a compulsory constituent of the B.Sc.(Econ.) and B.Com. degrees taken by the vast majority of L.S.E. students. In the B.Sc.(Econ.) there was a choice of 'special subjects', one of which was geography, so from 1912 L.S.E. may be said to have pioneered an Honours degree in geography. Indeed, between 1921 and 1945 more than forty per cent of the 569 geography Honours graduates in the Joint School took the B.Sc.(Econ.) degree.

The middle 1920s brought the first major changes. In 1925, Mackinder, by then an important figure in the political world, resigned his academic post, and was succeeded by Professor Rodwell Jones and in 1926 Dr L. Dudley Stamp, a graduate in both geology and geography from King's College, who had been first an oil geologist in Burma and then Professor of Geology and Geography in the University of Rangoon, joined L.S.E. as the Sir Ernest Cassel Reader in Economic Geography. In King's College Dr S. W. Wooldridge, assistant in Geology, was promoted to Lecturer in Geography in 1927; he became increasingly involved in teaching the physical basis of geography, including meteorology and climatology, and his intensive field studies in geology and landforms led him to develop an interest in the effect of the physical environment on the development of human settlement and land use, i.e. in historical geography.

Towards the end of the twenties there were two staff replacements at L.S.E. – W. G. East *vice* L. G. Robinson (for historical geography) and S. H. Beaver *vice* Miss W. Hunt (for all the practical 'map classes' and a variety of other teaching duties); while at King's College, Dr H. J. Wood (an L.S.E. graduate) was appointed as the first 'lecturer in geography'. There, with no further changes except at the research assistant-cum-demonstrator level, the staff situation remained for the decade of the thirties – an example of stability that was unparalleled in any other university at the time, and which most certainly could not be matched anywhere during the period since the Second World War.

One unique feature of the L.S.E. half of the Joint School was the evening teaching. All lectures and classes given in the day-time had to be repeated for the students who (at reduced fees and taking degree courses lasting for five rather than three years) represented a substantial part of the student total. In the years between 1923 and 1939, from thirty-seven to forty-five per cent of all L.S.E. students attended only in the evenings. This represented a burden on the staff that was recognized by the rather lighter day-time teaching loads than were common in other university institutions.

The outbreak of war in 1939 brought an abrupt end to this period of stability, and for the next few years the Joint School was distinctly disjointed. The L.S.E. was evacuated to Cambridge, with the administration focused on Peterhouse and the Department of Geography finding a welcome in the School of Geography in Downing Place. During the first year of the war the evening students were taught in London by Stamp and Ormsby – at premises outside the School, since it had been taken over by the Ministry of Economic Warfare. King's went to Bristol, and the separation was complete, apart from a joint field class held at Bradford-on-Avon in 1940, and some lecturing visits by Rodwell Jones to Bristol. Harassed by the 'blitz', King's returned to London in 1943, but L.S.E. was to remain in Cambridge until the end of the war. Dr Ormsby had retired in 1940, though she continued to make the journey to Cambridge to give occasional lectures until 1942. Rodwell Jones retired in 1945, to be succeeded in the Chair by Dudley Stamp, and Wooldridge, after a short period as Professor of Geography at Birkbeck College, was appointed as the first Professor of Geography at King's in 1947. Later, with Stamp's translation into a newly established Chair of Social Geography at L.S.E. in 1948, R. O. Buchanan (a graduate of the School) was appointed from University College to the vacant Headship of the Department, while East left L.S.E. to replace Wooldridge at Birkbeck.

The contribution of the Joint School to the growth of British geographical ideas and methodology between 1922 and the end of the Second World War may be analysed under a number of headings:

(i) *Economic geography* – especially the work of Rodwell Jones, Bryan, Stamp and Beaver;

(ii) *Regional geography* – major published studies of North America, Asia and France, with specialist teaching of Germany and the British Isles;

(iii) *Geomorphology* – of which Wooldridge was one of the greatest British exponents;

(iv) *Historical geography* – with the broad study of Europe by East and the detailed studies of Anglo-Saxon England by Wooldridge;

(v) *Land use and planning* – dominated by Stamp's Land Utilisation Survey, but with important contributions during the war by Beaver in the Ministry of Town and Country Planning, and by E. C. Willatts and several other Joint School graduates in research posts in that same Ministry, and by Stamp in the Ministry of Agriculture;

(vi) *School text-books* – in which field Stamp was virtually supreme in both Britain and many other countries;

(vii) Contributions to the work of other organizations, notably the Geographical Association, the Annual Conference of which was held in, and organized from the L.S.E. by Beaver for many years, and the Le Play Society, for which the Joint School – in the persons of Stamp, Ormsby and Beaver – provided leaders for field-study parties in many parts of Europe. It must also be remembered that members of the Joint School played a notable part in the foundation of the Institute of British Geographers. During the Second World War, Beaver, on secondment from L.S.E., made important contributions to the series of Naval Intelligence Handbooks issued between 1942 and 1947.

Economic geography

In the 1920s and 1930s economic geography was still largely descriptive and interpretative rather than theoretical. Wooldridge, in his Inaugural Lecture at Birkbeck College in 1945, expressed the opinion that 'it will require economists and geographers to build an economic geography worthy of the name', and a few years later (in Wooldridge and East, *The Spirit and Purpose of Geography*, 1951: 108) that 'economic geography as a specialist branch of geography has not yet advanced so far as to have established its position beyond question and to have differentiated itself clearly from other branches of the whole subject'. Be that as it may, there is no doubt about the value of the contributions slanted towards the interpretation of economic activity contained, for example, in Rodwell Jones's *North England: an economic geography* (1921; 2nd edn 1936); in the large economic section written by P. W. Bryan in Jones and Bryan's *North America* (1924; 7th edn 1946); in Stamp and Beaver's *The British Isles: a geographic and economic survey* (1933; 6th edn 1971); in Stamp's *Asia: an economic and regional geography* (1929; 11 further editions); and in Stamp's monumental revisions, dating from 1928, of G. G. Chisholm's *Handbook of Commercial Geography*. Some of these works were appreciated outside their country of origin. *The British Isles*, for example, was translated into Russian (quite without permission or payment, of course!), while *Asia*, that ran to twelve editions in English, was translated into Chinese and Spanish.

An early attempt at deducing 'principles' was Beaver's 'Localization of industry', a paper prepared for the International Geographical Congress held in Warsaw in 1934 and later published in *Geography* (20 (1935): 191–6); and his special interest in iron and steel produced not only two lengthy chapters in *The British Isles* but also a study of the iron-ore and smelting industries of the East Midlands (*Geography*, 18 (1933): 102–17).

Another special interest developed at L.S.E. was the geography of communications, especially railways; in part this arose from the evening classes provided for railway employees by Rodwell Jones, and later by Stamp and Beaver, and an interest in examining for the Institute of Transport, and in part from Beaver's own concern with all things relating to railway development and operation. Articles on the railways of Czechoslovakia (*Stephenson Locomotive Society's Journal*, 1932) and on the Bulgarian State Railways (*Railway Gazette*, 64 (1936): 1204–7) arose out of field-study in those countries, and were followed in 1941 by a major study of 'Railways in the Balkan peninsula' (*Geographical Journal*, 97: 273–94). An article on 'The railways of great cities (*Geography*, 22 (1937): 116–20) was an attempt to produce a 'model' (long before the modern era of model-building!) for the railway pattern of metropolitan cities, while 'Gradients of the London and Birmingham' (*Modern Transport*, 1938) was a study of the relationship between Britain's earliest main line and its geological environment. This interest in railways was put to good use during the war through the contribution on rail transport made by Beaver to the Naval Intelligence Handbooks (see below, p. 88).

A major work which combined aspects of transport, historical and economic geography was Rodwell Jones's *The Geography of London River* (1932). This was a study of the effects of the physical nature of the river's course and of changes in the technology of shipping on the trade and industry of the Port of London. Another work on transport geography was Ormsby's 'The Danube as a waterway' (*Scottish Geographical Magazine*, 39 (1923): 103–12).

Regional geography

Regional geography, widely regarded between the wars as the core of the subject, or even as its culmination, took two forms: the actual description and delimitation of regions (however defined) within continents or countries, and the general geographical analysis of the physical background and man's use of it in areas of continental or sub-continental scale or in individual countries. On the continental scale, Jones and Bryan's *North America* and Stamp's *Asia* have already been mentioned. Covering a smaller area, and thus more detailed, is Ormsby's *France* (1931); and still smaller, her *London on the Thames* (1924) and her essay on the London Basin in A. G. Ogilvie's *Great Britain: Essays in Regional Geography* (1928). Two studies of a specific type of region are Ormsby's 'The limestones of France and their influence on human geography' (*Geography* 17 (1932): 11–19) and Beaver's 'The Jurassic scarplands' (*Geography*, 16 (1931): 298–307). Stamp was in his element in devising and analysing 'natural regions' (of the Herbertson type), and his school texts, noted below, were generously illustrated with maps and descriptions thereof; an early example not in a textbook was his article on 'Natural Regions of India', (*Geography*, 14 (1928): 502–6); another is 'Suomi: Finland' (*Geography*, 16 (1931): 284–97) that arose out of a Le Play Society excursion, referred to on p. 88.

Much of the teaching in the L.S.E. section of the Joint School was essentially based on the importance accorded to regional geography between the wars – by the profession, by the Examination Boards, and by the 'man in the street'. In the 1930s candidates for the London B.A. and B.Sc. Honours degrees sat nine papers, of which five were 'regional'. There were lecture courses on the British Isles, Europe, France, Germany, North America, Asia and the 'Remaining Continents', with the emphasis in the last of these being placed on 'homoclimes', i.e. regions of broadly similar climates but quite different cultural backgrounds. Rodwell Jones's Inaugural Lecture (*Economica*, 1925: 241–57) contains this sentence (255): 'Specialization in the department is essential, and it is becoming increasingly recognized that such specialization must be on a regional basis.'

A view of regional geography from the south side of the Strand formed the subject of Wooldridge's Presidential Address to the Institute of British Geographers at its Cambridge meeting in 1950, and to this the reader is referred ('Reflections on regional geography in teaching and research', *Transactions of the Institute of British Geographers*, 16 (1952), 1–11).

Geomorphology

This branch of the subject – a development from what had been known in earlier years as either physical geography or physical geology, depending on one's background – was the province of the King's College half of the Joint School. Taught in the 1920s by Professor W. T. Gordon, it was increasingly taken over by Wooldridge, who was one of its greatest exponents in Britain during the 1930s and later, and who brought great academic distinction to the study of landscape evolution. From 1921 to 1929, apart from one or two excursions into igneous petrology, almost all of Wooldridge's publications were concerned with the rocks and landforms of the London Basin and the Weald – including such major reviews as 'The structural evolution of the London Basin' (*Proceedings of the Geologists' Association*, 37 (1926): 162–96) and 'The Pliocene history of the London Basin' (*Proceedings of the Geologists' Association*, 38 (1927): 49–132). These were followed in 1932 by a paper that showed his increasing interest in the geographical consequences of his detailed geological field work, 'The physiographic evolution of the London Basin' (*Geography*, 17: 99–116). Concentration on the teaching of geomorphology led in 1937 to his textbook (jointly with a junior author, R. S. Morgan) on *The Physical Basis of Geography: an outline of geomorphology* (1937); and further research, undertaken with his former pupil and colleague David Linton, led to the magisterial monograph, *Structure, Surface and Drainage in South-East England*, published by the Institute of British Geographers in 1939. Finally, his views on the development of the subject were lucidly set forth in his Inaugural Lecture delivered after his return to King's College as Professor of Geography in 1948 on 'The role and relations of geomorphology'. But Wooldridge's wide reading and ready pen also led him into the fields of historical geography, land utilization and land-use planning (see below).

Historical geography

There were interests in historical geography on both sides of the Strand, with A. P. Newton, Professor of Imperial History, and Dr H. J. Wood, specializing in the history of geographical discovery, at King's, and W. G. East teaching the historical geography of Britain and Europe at L.S.E. The historical tradition at L.S.E. stemmed from the interests and writings of Sir Halford Mackinder, whose *Britain and the British Seas* (1902) was a pioneer work in the period preceding the First World War, and whose magnificent volume on *The Rhine* (1908) was essentially a study in historical geography. Rodwell Jones, too, was an able exponent of the art of seeing the geographical influence upon the course of history, as Part I (134 pages) of Jones and Bryan's *North America* (which was adapted from his doctoral thesis) bears witness. Indeed, Rodwell Jones made much of historical geography in his Inaugural Lecture at L.S.E. in October 1925 (*Economica* (1925): 241–57). 'Here lies', he said, 'the greatest field for research and the most cultural part of the subject' (255); and, 'if geography has come to mean a study of man's relation with his physical environment, then all geography is historical geography, just as all subjects of study are a part of history or, at least, have their historical aspect' (250–1).

It was, however, East who played a major role in the establishment of historical geography as a worthy academic discipline. There was some debate in the 1930s as to the methodology of the subject, which seemed to resolve into two alternatives, either the 'reconstruction of past geographies', yielding a series of cross-sections or 'period-pictures', or the analysis of geographical influences on the course of history at different periods and in different regions or localities. East's *Historical Geography of Europe* (1935) was a blend of both types, with chapters on subjects such as 'Rural settlement and agriculture', 'The Byzantine Empire' and 'Europe in the Railway Age'. H. C. Darby's collection of essays on *The Historical Geography of England before 1800* (1936), to which East contributed a chapter on 'England in the eighteenth century', was a collection of 'period-pictures'. East's other major work, *The Geography behind History* (1938) was based on a series of themes, such as 'routes', 'towns', 'frontiers and boundaries', by which the importance of geography as an historical influence was demonstrated. More detailed studies were made of two English ports, Kingston-upon-Hull (in *Economica*, 11 (1931): 190–212) and Whitby (*Geographical Journal*, 80 (1932): 484–97); and another thematic essay appeared on 'Land utilization in England at the end of the eighteenth century' (*Geographical Journal*, 89 (1937): 156–72).

Land use and planning

The work that professional geographers, and many others, would most readily associate with the Joint School in the 1930s is the Land Utilisation Survey of Great Britain, organized by L. Dudley Stamp, housed for most of its life within L.S.E., and to which so many former students of the Joint School made noteworthy contributions.

The history of the Survey has been very fully documented by Stamp in Chapter I of his monumental *The Land of Britain: its use and misuse* (1948), and it is unnecessary to repeat it here. The idea of making field-by-field surveys of small local areas had occurred to many people in schools and in university departments of geography (see, for example, 'Regional surveys', in *Geographical Teacher*, 13 (1926): 362–7, and W. S. Baker, 'Some notes on a regional survey', *Geographical Teacher*, 13 (1926): 451–3). In L.S.E. student exercises on land-use mapping were a regular feature of the teaching, and one example, from Surrey, was illustrated by Rodwell Jones in an article on 'Commodity maps' (*Economica*, 1 (1921): 253). This paper also contains an almost prophetic suggestion: 'If local schools of adjacent districts were to undertake such work in a systematic way, useful combined maps of very considerable areas could be easily produced and reduced. County education authorities might well further such a scheme' (252). These early efforts at land-use mapping culminated in just such a whole-county exercise organized in Northamptonshire schools and published in three one-inch sheets by the Ordnance Survey in 1929. It was this, more than anything else, that inspired Stamp to organize a more detailed survey covering all the counties of England, Wales and Scotland, on a uniform basis of classification. The survey work was to be carried out on six-inch maps, with eventual publication on the standard one-inch sheets of the Ordnance Survey.

The work of organizing the surveying of some 22,000 six-inch quarter-sheets, of reducing the results to the one-inch scale, of encouraging authors to write county reports and, above all, of financing the whole operation was colossal, and it may be seriously doubted whether there was anyone other than Stamp in the profession in the 1930s who would have had the energy, the drive and, above all, the business acumen to carry the task through. Many of the Directors of Education in the counties played an important part in the organization of the survey work through the schools, and county 'patrons' were tapped for contributions to the expenses. But Stamp himself made the largest contribution to the costs of the Survey and there were substantial grants from the Rockefeller Foundation and the Pilgrim Trust.

The Organizing Secretary of the Survey, almost from the beginning, was Dr E. C. Willatts, a graduate of L.S.E. As the years went by, more and more of the Joint School students, graduates, current and former members of the staff came to be involved. Thus the Yorkshire, West Riding, Report, written by S. H. Beaver, records that nearly 200 of the 553 six-inch sheets covering that large county were surveyed, mostly around 1937, by students or ex-students of the Joint School; and of the ninety-two County Reports no less than fifty-six were written wholly by members or ex-members of the Joint School, with ten other Reports to which contributions were made. In truth this is a remarkable record. Amongst the contemporary staff of the Joint School, the names of Stamp, Wooldridge and Beaver appear on the list of authors – Stamp himself wrote twelve Reports and contributed to five others – and other graduate authors who subsequently achieved professorial status in other places include D. W. Fryer, F. K. Hare, J. H. G. Lebon and A. C. O'Dell.

The Land Utilisation Survey did not end with the publication of the one-inch maps and county reports. Before these two aims were achieved (the last County Report was published in 1946) the outbreak of war in 1939 made the Survey of immediate and practical value to the nation as the basis for planning the 'plough-up' campaign, and subsequently for much further research and publication by the Ministry of Agriculture and the Ministry of Town and Country Planning. Just before the war, however, the Royal Commission on the Distribution of the Industrial Population (the Barlow Commission) requested the Survey to prepare a Land Classification map, and a preliminary bulletin, entitled 'Fertility, productivity and classification of land', with three maps (of good quality, poor quality and intermediate quality land), was published in 1941. It was of immediate interest to another Commission – the Scott Committee on Land Utilisation in Rural Areas (of which Stamp was Vice-Chairman), and a Land Classification map, showing a ten-fold classification, was made for its Report in 1942; subsequently a 'ten-mile' (actually 1:625,000) map was published by the Ordnance Survey.

The culmination of the Survey was the publication in 1948 of *The Land of Britain: its use and misuse*, Stamp's distillation of the essence of the county reports. It demonstrates, in Buchanan's words, 'the trend of Stamp's thinking, based on the record of how the land was in fact being used, but concerned progressively with how it should be used'; and it earned for him the first-ever D.Lit. degree of the University of London in Geography.

In 1942 Stamp was appointed Chief Adviser to the Minister of Agriculture in Land Utilisation (and his services were recognized by the award of the C.B.E. in 1946). In the following year came the formation of the Ministry of Town and Country Planning, with Willatts in charge of the Maps Office, and Beaver as Research Officer responsible for matters involving geology and the working of minerals. Other Joint School graduates holding research posts in the Planning Ministry at one time or another in the 1940s included F. H. W. Green (at Bristol, and later in London), G. M. Hines (at Leeds), J. Stephenson (at Cambridge), G. J. Husted (at Reading), J. R. James (at Newcastle), and S. W. E. Vince

(London), while A. C. O'Dell occupied a similar post in the Planning Division of the Scottish Department of Health.

Beaver's contribution in the Ministry included a series of monographs (never published, unfortunately) on all the country's surface-worked minerals and the industries arising therefrom; but a major work was published, *Derelict Land in the Black Country* (1946). This provided the background for the extensive programme of reclamation that followed, and was itself succeeded by similar surveys in the Potteries and Shropshire coalfields.

Finally, in the planning field, the Waters Committee on the sand and gravel industry, that sat from 1946 to 1954 and published eighteen volumes of reports, included Wooldridge and Beaver in its membership. Though generally regarded by their colleagues as geologists, their contribution to the work of the Committee was very much coloured by a geographical viewpoint that combined an appreciation of both space and economics.

School text-books

Apart from an early work by Beaver (*The Americas*, published in 1932, in a series edited by Dr Marion Newbigin, by Collins), this field was monopolized by Stamp between 1924 and 1949. Beginning in Burma, and arising from his residence there, with two books written jointly with F. G. French, *A Geography of Burma for Schools* (with a Burmese translation) and *The Indian Empire*, in 1924–26, Stamp's return to England as the Sir Ernest Cassel Reader in Economic Geography at L.S.E. was marked by the appearance in 1927 of *An Intermediate Commercial Geography*, Part I, followed a year later by Part II; these were written as texts for the Intermediate B.Sc.(Econ.) and Intermediate B.Com. examinations of the University of London. Also in 1927 came *The World: a general geography for Indian schools*, which was translated into seven Indian languages. A South African edition followed in 1927, an Australian in 1928, and in 1929 a version for British schools that before its author's death in 1966 had run into seventeen editions. At a lower academic level the *New Age Geographies*, in four books, were written jointly with his wife, Mrs Elsa Stamp (who was herself a graduate of King's College), while for Intermediate studies (in the University of London sense, that is, post-matriculation and pre-degree) *A Regional Geography*, in five volumes, saw the light of day in 1930; Part II, *Africa*, was delayed until 1934, and was written largely by Beaver. These five books each achieved between seven and fifteen editions.

Designed particularly for candidates for the Institute of Bankers examinations (for which Stamp was chief examiner in Geography for many years), *Commercial Geography* was published in 1936, with six subsequent revised editions. Then, as if to demonstrate his versatility, he reverted to earlier interests with *Physical Geography and Geology* in 1938, with a specially modified edition for use in India in 1939.

Concurrently with these, a further series of school texts was being produced, *Geography for Today* (1937–9) in four books for School Certificate studies, prepared under Stamp's guidance by a small committee that included his colleague Beaver; with the help of G. H. T. Kimble (a Joint School graduate) this series was adapted for use in Canada (in 1939–40), and some years later (1949), with the assistance of G. E. D. Lewis (another Joint School graduate), a Malayan version was produced.

Meanwhile, several individual books and series were produced for Indian schools, in English and numerous languages of the sub-continent, and there can be few children who passed through the Indian educational system in the 1930s and 1940s – and perhaps also for the ensuing two decades – whose knowledge of geography was not derived from Stamp.

This spate of text-book publishing may have done little to advance the reputation of British 'academic' geography; but it provided a sound background of facts and ideas upon which undergraduates in universities could build their degree studies, and it extended the influence of British geographical teaching to most parts of the English-speaking world and, through translations, to many other areas as well.

Contributions to other organizations

The Geographical Association During the 1920s and 1930s and, after a war-time interruption, in the late 1940s, the Annual Conference of the Geographical Association was held at L.S.E. and from 1933 to 1950 was organized by Beaver. Although membership of the Association consisted in the main of practising school-teachers, the Annual Conference always attracted a sprinkling of university geographers because of its attractive programme of speakers and its Publishers' Exhibition. The Joint School Branch – commonly and affectionately known as the 'Geog. Ass.' – was one of the largest and most active in the whole Association.

The Le Play Society This 1930 offshoot of Le Play House (the Sociological Society), under the Presidency of Sir Patrick Geddes and with the help of Professor C. B. Fawcett of University College London, had as its main object the furtherance of geographically based sociological studies of localities and communities in Britain and Europe (see Beaver, 'The Le Play Society and field work', *Geography*, 47 (1963): 225–40). Several members of the Joint School staff made notable contributions to its programme of fieldwork and publication: Stamp led excursions to Finland, Yugoslavia and Portugal, Dr Ormsby to Romania, and Beaver to Bulgaria and Albania. Publications resulting from these studies are listed in Beaver (1963). This foreign fieldwork was in fact simply carrying on a tradition that had become established in the Joint School Branch of the Geographical Association, for which members of the staff led excursions in the late 1920s and early 1930s to Belgium, Switzerland, Germany, Czechoslovakia, Greece and Finland.

The Institute of British Geographers The inaugural meeting of London geographers in 1931 that led ultimately to the foundation of the Institute of British Geographers was attended by Wooldridge, East, Wood and Beaver from the Joint School (see *Transactions of the Institute of British Geographers*, 20 (1954): 3; R. W. Steel, *The Institute of British Geographers: the first fifty years*, 1984). At the first General Meeting of the Institute in January 1933, Beaver read a paper on the 'Relative development of the iron industry in Northamptonshire and Lorraine'; at the third, in 1935, Wooldridge contributed a paper on 'Facets as ultimate units of geographical analysis' (this, unfortunately, never appeared in print); and at the fourth, in 1936, Beaver and East were on the list of contributors. All these early meetings were held at L.S.E., concurrently with the Annual Conference of the Geographical Association.

War-time activities In so far as they concerned land use and planning, these have been mentioned above. Another activity that claimed the attention of many geographers, including several Joint School staff and former students, was the production of the series of Naval Intelligence Handbooks. At the Cambridge centre Beaver was a member of the full-time staff from 1941 to 1943. He was concerned mainly with the analysis of the railway systems in France, Belgium, Denmark, Germany, Greece and Yugoslavia, and with the writing and editing of the handbook of the Netherlands East Indies. Other members of the Joint School who contributed to the sixty or so volumes that were eventually issued included Wooldridge, Ormsby and O'Dell. Two other members of the staff who took war-time posts were Wood (in the Admiralty) and East (in the Ministry of Economic Warfare).

Conclusion

This essay has attempted to assess the significance of the role played by the Joint School of Geography in the London School of Economics and King's College, London, in the years between the wars. In many of its activities – for example, the establishment of the Institute of British Geographers – the Joint School was closely associated with departments of geography elsewhere; and the University of London played an important part in the academic development of the subject. As with many other groups of geographers between the wars, numbers were very small compared with the situation today. This underlines still more the importance of the contribution made by the staff – and the students, too – of the Joint School to the laying of the foundations and the building up of inter-war geography from which so much of the subsequent development of the subject has stemmed.

NOTE

Much of the material in this essay derives from the personal recollections of the author. For additional information, see the Introduction to the Rodwell Jones Memorial Volume, *London Essays in Geography* (1951) (L. D. Stamp and S. W. Wooldridge eds), and three short papers in *Horizon*, 22 (1973) (an issue to celebrate the golden jubilee of the Joint School): 'The Joint School: early days' by W. G. East; 'The virtues of requisite variety' by F. K. Hare; and 'A has-been looks back' by R. O. Buchanan.

7 Geography in a University College (Nottingham)

K. C. EDWARDS*

After winning a borough scholarship in 1922 I entered the University College of Southampton. In recent years Southampton had had two well-known men in charge of geography: C. B. Fawcett (1915–19), appointed to a newly created lectureship in the subject, and W. H. Barker (1919–22) who was given a Chair in 1921. As in many other universities at that time they were not trained in geography but came from some other subject or subjects in which they had specialized. Their skills in these rendered their work for geography as varied as their initial training but at least they established a common mode of approach which in turn drew much of its inspiration from the new geographical studies which emerged from Oxford or from the work of such writers as H. R. Mill of the Royal Geographical Society. Fawcett, who had taken a degree in science at the University College of Nottingham, went on to extend his qualifications by taking the invaluable geographical Diploma at Oxford under A. J. Herbertson. Thus he was a singularly well-qualified person for appointment to the new lectureship. In 1918 he published *Frontiers: a study in political geography* and in the following year there appeared his *Provinces of England*.

Another advantage for the geography course at Southampton was the presence in the town of the headquarters of the Ordnance Survey, at that time under the direction of Colonel Sir Charles F. Close. In my second year in the Department of Geography the Director-General became well disposed towards the students and allowed us many privileges.

* Kenneth Charles Edwards, C.B.E. (b. 2 March 1904) studied geography in University College, Southampton, and was awarded a B.A. Honours degree of the University of London in 1925. His subsequent career was spent entirely in the University of Nottingham, where he was appointed to an assistant lecturer in geography in 1926 and subsequently as the foundation Professor in 1948, a post he held until his retirement and election as Emeritus Professor in 1970. He died in Nottingham on 7 May 1982.

One overriding difficulty of the degrees given by the University Colleges of this period – Exeter, Hull, Leicester, Nottingham, Reading (until it became a full University in 1926) and Southampton – was that they were bound to accept the external syllabuses of the University of London. These were virtually the same as those of the Internal Colleges of London which were generally larger institutions with greater scope for training and research. The examinations, including the Intermediate (a first-year four-subject post-matriculation stage), were all literally external, the candidates being judged by examiners appointed by London, with no contact between them and the students' own teaching staff. To this regulation there was no exception and this situation presented many difficulties. Above all, it was impossible to maintain a full range of the branches of geography which, following the First World War, were developing freely. Moreover unsatisfactory stipends prevented the University Colleges in particular from employing more than one or two members of staff.

I had to exercise a careful selection of subjects for my course. Geography was in the Faculty of Arts (only later was it available as a Science), but one science was allowed. In the Faculty of Arts all students had to pass in Latin for the Intermediate stage. In addition to geography and Latin, I chose English, in which I had been especially strong at my school. My fourth choice was geology which was clearly a cognate subject and allied to my main quest.

The Department of Geography was in the hands of a Reader, O. H. T. Rishbeth, who had come from South Australia with a first degree from Adelaide, but who had gone to Merton College, Oxford, to extend his classical education. Rishbeth wished to teach geography, however, and he was appointed to Southampton after a year's initial experience under Professor H. J. Fleure at Aberystwyth. He was a tall individual, somewhat aristocratic in manner but kindly and generous to students. He was hardly a good teacher but had an excellent style as a lecturer and performed well with the carefully prepared topics which he handled. He was especially good at German and with the requisite textbooks in French allocated to his students, they were well favoured in the use of foreign languages. His lectures on the origin and structure of the earth were traced from J. L. Kober's *Der Bau der Erde* which was more recent and more detailed than anything published in English. Much of this work was translated for us by Rishbeth himself.

Rishbeth did well in re-establishing his Department as a new force in the quest for geography and, though it was delayed until the year following my graduation, he received his Chair in 1926. He had one colleague, Miss F. C. Miller, who shared the teaching load and did all the laboratory work. She was the authority on all questions relating to maps and was responsible for cartography, for historical and political geography, and for various regional courses.

Two small books were recommended to students in their first year: F. S. Marvin's *The Living Past* (1913), a sketch of Western progress, and *The Dawn of History* by J. L. Myres (1911), a similar but more detailed treatment. Both books were historical but the significance of the geographical background was emphasized. For the Intermediate examination the book recommended was *General and Regional Geography* by J. F. Unstead and E. G. R. Taylor.

A booklet about the post-war settlement was M. I. Newbigin's *Aftermath: a geographical study of the Peace Terms*, published in 1920. This turned our attention to a new Europe, and indeed to a new international order, to which the League of Nations offered its guidance. But the study of Europe presented students with a very real problem, largely of a philosophical nature. There were two quite different advanced textbooks on this continent. One was L. W. Lyde's *Continent of Europe* which had appeared in 1913. This was a detailed study of the countries involved but it was a difficult book to understand, partly because of the author's complex style, and partly because as a writer he examined countries from the political standpoint and not on the basis of Herbertson's concept of natural regions. It was only partially revised in 1924 which meant that in many respects it was out-of-date. The other book, *Europe: a regional geography* by Nora E. MacMunn and Geraldine Costar, had been planned by A. J. Herbertson before his death in 1915 and broadly followed the lines of his famous paper written ten years earlier ('The major natural regions: an essay in systematic geography', *Geographical Journal*, 25 (1905): 300–12). It was given to two of his pupils to write, with his own introductory chapters, to provide a geography which should be regional 'in a true and logical sense of the word' for training colleges and the upper forms of schools. At the Intermediate stage in the London degree course this helped to bring students up-to-date with the new methodology transmitted by Herbertson. But for advanced work Lyde proved the more valuable, despite the above-mentioned difficulties.

As a graduate in geography I spent the last year of my work at Southampton following the postgraduate course of training for teachers (University of Cambridge External). In that year, together with the following long vacation, I learned what was meant by research, for I gladly helped Rishbeth prepare the way for his work on the Tertiary and Pleistocene features north of Southampton. Thus I became a trained geographer and, like others of my generation, I had chosen the subject from the beginning and helped to establish it on its upward growth.

Who was the leading geographer of the day during my time as an undergraduate reading geography? This was an important question to a student then in the throes of learning the names of many individuals, and my choice fell upon Dr Marion I. Newbigin. She did a great deal for geography in Britain and seems to me to be the natural successor to the Oxford group led by Mackinder and Herbertson. She was trained as a biologist in Edinburgh, hence her supreme value to geographers as a scientist. Yet she never held a professorship, though she gained a D.Sc. from the University of London and later on was appointed as an external examiner for the same University.

Dr Newbigin was appointed to the editorship of the *Scottish Geographical Magazine* which, in her hands, with meticulous care, became the most expressive journal in the country and had for many years the most influential position among geographical periodicals in Britain. This task gave her a remarkable insight into the detail of many developments in geography.

All her books were written in a lucid and refreshing style. Well before the First World War, she published in the Home University Library a volume entitled *Modern Geography* in which the world was considered as the home of man and his activities. The book was so successful that it lasted for many years. In 1912 (with a reprint in 1921) came an *Introduction to Physical Geography* and in the same year *Man and his Conquest of Nature* which thus became one of the earliest works on human geography together with A. J. and F. D. Herbertson's experimental junior book written ten years previously. She recognized the basic needs of good clothing and shelter in human communities and showed how the spread of mankind was blessed by nature in some environments and not in others. She was familiar with the French views of F. Le Play, Vidal de la Blache, and above all Jean Brunhes, and saw clearly what was meant by the term human geography. Although it was not original, the book guided the way for later teachers such as H. J. Fleure and P. M. Roxby.

In 1913 Dr Newbigin produced both *Ordnance Survey Maps: their meaning and use* and *Animal Geography*, which was based on her teaching as an extra-mural lecturer for the Edinburgh Medical School. It is probable that she might also have written a geography of plants but this was undertaken by Dr M. Hardy, though unfortunately there was a long delay before it was published in 1920. In 1916 came her scholarly book on the *Geographical Aspects of the Balkan Problem*. In this she owed much to the earlier work of J. Cvijić and in it she called attention to the changing run of the Adriatic coastline in the neighbourhood of Scutari. When Cvijić published his detailed work, *La Péninsule Balkanique*, in Paris in 1918, he noted this query and in dealing with it (on page 24), acknowledged it with a footnote.

In 1920 came the *British Empire beyond the Seas* which, following Herbertson, was built upon a framework of seasonal climatic phenomena. Two years later there appeared *Frequented Ways*, a travel book on European journeys with the geology, climate and vegetation as fundamental guides. Not surprisingly Dr Newbigin also did what many others had done, she wrote a study of the Mediterranean lands, but she did so from the powerful vision of human and historical geography and as a result the book, *The Mediterranean Lands*, published in 1924, was an outstanding success. She endowed the subject with its scientific character in a manner which no one else had done so positively before; she recognized clearly what was meant by the term human geography; and for more than thirty years she rendered outstanding service to geographers through the *Scottish Geographical Magazine*, by far the most impressive periodical in Britain throughout her editorship.

In 1926 I was appointed to the Department of Geology and Geography under Professor H. H. Swinnerton at the University College of Nottingham. Swinnerton was trained initially in the fields of zoology and geology and came to Nottingham in 1902 with a D.Sc. in zoology. His early publications showed how he changed the emphasis, as time went on, from zoology to geology, incorporating also a knowledge of botany. In 1911 he was given the Chair of Geology and in due course produced his epoch-making *Outlines of Palaeontology* (1923). During the twenties there was a steady rise in the number of students wanting to read geography in many British universities. At Nottingham they were all catered for by Professor Swinnerton with another lecturer, C. G. Beasley, who came from University College London.

My own post was something of an experiment, since I was an assistant-lecturer in both sections of the Department, which at that time was a very large one. Besides the degree courses in both geology and geography (most students of the latter were persuaded to take geology as their subsidiary subject), there was a large two-year course for the training of elementary teachers, a course taken by the novelist, D. H. Lawrence, instead of a B.A. degree, between 1909 and 1911. He was expert at the drawing of flowers so that Swinnerton, his botany lecturer, reported favourably upon him. Lawrence was not only born nearby at Eastwood but was already writing *The White Peacock* and *Sons and Lovers*, both stories of his home district. In addition there were students of the Mining Department who did practical geology twice a week in the laboratory.

Part-time degree students were admitted on Saturday mornings and there were two or three graduates working for their M.A. or Ph.D. by research. The working week was heavily loaded with lectures or laboratories so that in the first few years of my time in Nottingham I found myself doing between twenty-eight and thirty hours each week.

In 1930 Beasley resigned to go to Rangoon to the chair of geography which had been occupied until 1926 by L. D. Stamp. Swinnerton, acting on my advice, appointed Neville V. Scarfe, another product of University College London. We became good friends, saw eye to eye on the progress of geography, and shared fully in the affairs of staff and students in the College.

In the following year I took the London M.A. (External). A dissertation on a subject of my own choice and two written papers were the requirements. This was not an easy degree to obtain. There were six candidates, and only myself and one other passed.

In 1932 Scarfe and I invited Professor Fleure to talk to the local branch of the Geographical Association. In the following year we obtained the help of the Public Lectures Committee to invite Sir Halford Mackinder to speak to all the teachers of geography in the district. There was an audience of over four hundred who heard him give an inspiring address on 'the geographical way of thinking'.

After another two years Swinnerton, who was nearing sixty, and was fully involved with the usual crop of geology graduates and undergraduates, thought that a separate Department might be founded for geography. He wisely made one or two stipulations. Thus he would continue to cover the physical basis of geography, including climatology and, if there was a demand, plant and animal geography, while I was to maintain, as far as I could, the intake of subsidiary geologists. This arrangement was accepted by the Senate and, without discussion, I was advanced to the grade of Lecturer and became the Independent Head of the Geography Department. This I achieved at the age of thirty and I was very grateful to Swinnerton for his generosity and goodwill.

At about this time I became a founder member of an entirely new group convened in London – the Institute of British Geographers, a body of professional geographers who came together for purposes of research and publication. The channels for the latter were quite inadequate, and how needy they were later results have shown. At the third meeting of the Institute I gave a paper on the 'Modern waterway of the River Trent'.

An initial success of the new Department of Geography was the achievement of the very able student, K. B. Cumberland, who took the advice of R. O. Buchanan (then a lecturer at University College London) and went to New Zealand. After a few years in Christchurch, he became the first Professor of Geography in Auckland in 1943. Another graduate, James Fox, followed him to Auckland after the war and, after a prolonged sojourn there, moved to Australia to the University of New England at Armidale, New South Wales, where he was appointed to a Chair in Human Geography. A. G. Powell, another graduate, followed me into the Ministry of Town and Country Planning and eventually rose to be Chief Planner of the London and South-East Region and an Under-Secretary of State.

Fieldwork

The fully trained geographer must have a realistic model which he knows, and Swinnerton and I took our degree students away every year during the Easter vacation for this purpose. It was much more difficult to take them abroad, and I was glad to have an opportunity to encourage this through my association with the Student Group of the Sociological Society.

Few geographers of my age failed to hear of Patrick Geddes (1854–1932) who was a biologist from Dundee and Edinburgh. He was a far-seeing man, full of ideas and stimulating thought. In addition to his scientific training he became a sociologist and ultimately a city planner. He was fluent in French, loved France, and knew a number of French scholars including the geographer E. Reclus. Geddes discovered Frederick Le Play, who lived half a generation before Vidal de la Blache, and wrote *Les ouvriers européens* in 1879. He was a mining engineer, not a geographer, but his work as a sociologist left an indelible impression upon Geddes and through him upon countless geographers in their training early this century. Le Play travelled throughout the European countryside seeking out types of settlement, using the family budget as his measuring rod, the sociological test. He grouped his observations according to the trilogy: place–work–folk.

Geddes and others founded the Sociological Society in 1904 and after 1920 it met at Le Play House in London where Miss Margaret Tatton was in charge of its educational tours. I went in 1928 to Aldrans in the Austrian Tyrol with the student group but the fieldwork was wholly sociological and the few geographers present did less than justice to the objective. After our return I pleaded with Miss Tatton to give equal opportunity to geographical surveys of a 'regional' character for the subject was now booming and there was no other way of working abroad at so low a cost. I won my point, and with myself elected as organizer, the student group changed its nature.

Later, after visiting Luxembourg in 1932, I found a major interest in that country. During the Second World War I was responsible for the Admiralty (Naval Intelligence Division) handbook on Luxembourg (published 1944) and later I chose some geographical aspects of that small country for my Ph.D. Among other publications I produced the sheets of the first atlas of the Grand Duchy with the assistance of several Luxembourg scholars. In recognition of this work I was awarded the Order of Merit of the Grand Duchy.

In 1934 and 1935 there were changes in the organization of the Sociological Society in London. The Society withdrew from Le Play House and the latter became known as the Le Play Society and lasted until some years after the Second World War. (See S. H. Beaver (1962), 'The Le Play Society and field work', *Geography* (47: 225–40)). Throughout this period there were students from the Nottingham Department of Geography who were members.

Following the war a new venture, much like the old, was started. This was the Geographical Field Group (G.F.G.) which was in a sense the climax of all the preceding events. The aim was as before but absolute freedom of action was invested in it. The headquarters were in the Department of Geography at Nottingham and I was elected its first President and remained in that office for many years.

Nowadays many university students undertake foreign travel as part of their courses but many Heads of Departments in addition favour the help and goodwill of the G.F.G. The University of London periodically circulates information about the G.F.G. to its external students. Further help has been gained by an occasional grant from the Soddy Trust, a relict of Miss Tatton's days with the Le Play Society.

Regional planning

As a geographer interested in planning, I was asked by Professor E. G. R. Taylor in 1938 to take part in the collection of data which led to her evidence for the Barlow Report on the 'Distribution of the Industrial Population' (Cmd. 6153, 1940).

In 1943 I was approached by the newly created Ministry of Town and Country Planning to take charge of research in my own region. There was to be a Regional Research Officer, a full-time post, for each of the ten Standard Regions with a number of others at the Ministry's headquarters. In Nottingham I was second-in-command under the Regional Planning Officer who was an architect from Sussex. He was without any knowledge of industrial areas, of coalfields, and of problems of mining subsidence, all of which I knew at first hand. On the whole we worked reasonably well together.

Just after my appointment I met John Dower, an amenity expert at headquarters with S. H. Beaver who was studying mineral working. We traversed together the Tunstead Limestone Quarry and the Hope Cement Works. I was greatly impressed by the personality of Dower, who at the time was preparing a report on the creation of National Parks. The Peak District, which particularly interested me, was scheduled, along with the Lake District and Snowdonia, in the Hobhouse Report (with John Dower as a member) in 1950.

Other events in the years immediately after the end of the Second World War are strictly beyond the scope of this essay, but they resulted directly from the development of geography between the wars. There was, for example, the Schuster Report on the Qualification of Planners, published in 1950. The Committee felt that the Barlow, Scott and Uthwatt Reports had established the legislative machinery and that, as the public had become genuinely interested in planning, there was a need for a statement about the qualifications of planners. Among institutions submitting evidence were ten universities, of which Nottingham was one. Here was a marvellous chance for geographers to place themselves and their subject in the national light. I did just this, and it was gratifying that the Report recommended that the education of planners should be a university degree course particularly in a relevant subject such as architecture, economics, geography and sociology. This should be followed by a postgraduate diploma course in town and country planning at a recognized university school of planning.

The University College of Nottingham, to which I returned in 1946, after refusing, despite a much higher salary, the offer of the Ministry to give me a permanent post, became the University of Nottingham in 1948. In the following year I became the holder of the first Chair of Geography in the University. The staff consisted of three full-time members and a demonstrator, and there were about 150 students. As soon as possible, measures were taken to equip another more serviceable building. After another eight years, I was fortunate enough to plan, with a sympathetic architect, a huge building with all amenities, as a Faculty of Social Science where geography had its own rightful place. I was at that time Dean of the Law and Social Sciences Faculty.

My interest in planning continued throughout my career at Nottingham. I supported without hesitation the appointment of a Professor of Architecture in 1960 and helped to create in 1969 the Institute of Planning Studies with its own Professor within the Faculty of Social Sciences.

Another activity that extended from before the Second World War and continued for many years afterwards was linked to the Geographical Association, to which, in common with most academic geographers, I gave regular support. It was our own organization, consisting of teachers, professional men and women, and laymen. I responded to its needs whenever I could and in about 1935 I became President of the Nottingham branch, a position that I still held in 1982. In the post-war period I spent twenty years editing a series of booklets, *British Landscapes through Maps*, designed for teachers and students. In 1963 I served as the Association's national President and in 1973 I was invited to give the Herbertson Memorial Lecture. This lecture, instituted by the Council before the end of the First World War in 1917, and only two years after Herbertson's death, has been given at irregular intervals since then. I took as my title 'Sixty years after Herbertson: the advance of geography as a spatial science' (*Geography*, 59: 1–9). I reminded my audience of what Sir Halford Mackinder had written in his obituary notice of A. J. Herbertson in 1915 – 'that the Geographical Association owes more to Herbertson than to any other man, with the possible exception of Professor Fleure'. Nearly six decades later it seemed appropriate on that occasion to underline how much modern British geography owes to the pioneers of the subject in the first half of the twentieth century.

8 Geographers and their involvement in planning

E. C. WILLATTS*

In 1946 in his inaugural address at the London School of Economics entitled 'Applied geography', L. Dudley Stamp remarked on 'the almost parallel careers of geography and town planning'. Throughout more than the first half of the period with which this volume is concerned this was literally true, for parallel lines do not meet. Both subjects were slowly developing but by the early thirties neither had reached the stage of being held in great respect by other disciplines, and planners certainly had produced very few publications of general interest. The numbers of professionals in both subjects were few indeed. The Town Planning Institute had been formed in 1913 by architects, engineers and surveyors and its members worked principally in the departments of local government concerned with such matters as roads and drains. It had only a few hundred members by the time the Institute of British Geographers was formed twenty years later with, initially, less than eighty members.

Those latter pioneers were almost exclusively engaged, and very heavily engaged, in university teaching, without the help of technicians. A high proportion of their students expected to become teachers of geography, an increasingly popular subject by then well established in the curriculum of secondary schools. I was such a student. I had no thoughts of ever becoming a planner. Indeed, like most geographers, I had never heard of planning. Nor were the early thirties a time when young graduates, particularly in geography, could expect to plan a career involving moving from one field to another. The international economic slump of 1931

* Edward Christie Willatts, O.B.E. (b. 4 July 1908) graduated (B.Sc.(Econ.)) at the London School of Economics in 1930. He was Organizing Secretary of the Land Utilisation Survey of Great Britain from 1931 to 1942, and was a lecturer in geography at Birkbeck College, University of London, between 1939 and 1942. He became a Senior Research Officer in the Planning Department of the Ministry of Works in 1941 and was Principal Planner in the Department of the Environment from 1948 until his retirement in 1973.

resulted in unemployment for graduates as well as others. With a geography degree and a teaching diploma I sought a post as a geography teacher but not until the autumn of that year was I offered a post in a reputable grammar school.

Concurrently a chance meeting with Dudley Stamp changed my plans. The post of Organizing Secretary of the Land Utilisation Survey of Britain was vacant and he offered it to me. It was temporary and unpensionable, and the salary was less than that of a teacher – but it seemed exciting. I accepted and so began ten years of rewarding work which was to have a very marked impact on physical planning. The story of the Survey's work, its struggles in an era when research grants were virtually unheard of, how it succeeded in harnessing the energies of countless volunteers, and its achievements in publishing a complete set of one-inch maps of England and Wales and of the populous parts of Scotland, with specimen sheets of highland areas, together with its series of ninety-two county reports, has been fully told in Stamp's major volume, *The Land of Britain; its use and misuse* (Stamp 1962). The Survey's mapping and research work inevitably drew attention to the use, and abuse, unconscious as that may have been, of our most fundamental finite resource. This was not at first foreseen and it was only after the commencement of the publication of the county reports in 1936 that its findings began to make much impact on the minds of workers in other fields. I was the only professional member of our small staff and we were too involved in the completion of our publication programme to be able to spare much effort to bring our work to the attention of organizations outside the academic field. In those days the same could be said of most other workers in geography and kindred subjects. Nor could we possibly have imagined that fifty years later H. C. Darby would write of the Survey that 'without doubt it was the greatest achievement of British Geography to date' (Darby 1983).

In the thirties geographers and planners had very little contact, and seemed to have little in common. Consider the state of the planner's art. The first Act of Parliament was passed in 1909 and was aimed at 'securing proper sanitary conditions, amenity and convenience in connection with the laying out and use of the land (being or likely to be developed for building) and of any neighbouring lands'. In the words of the President of the Local Government Board, it was enacted to improve the 'physical health, morals, character and the whole social condition of our people'. It applied only to new housing and it led to innumerable inter-war council housing estates, where happiness was represented by a subsidized three-bedroom house with a garden, while the private owner looked for a

£595 semi-detached house in a piece of ribbon development beside public services and a bus route.

In 1919 another Act, designed to stimulate post-war building, required planning schemes to be prepared for all towns of more than 20,000 inhabitants but in practice this was a failure and by 1932 ceased to be compulsory. In that year a further Act aimed at providing some controls. There was, however, little or no evidence of regard for social and economic objectives and ten years later a government committee declared that the Act had been 'a story of high hopes and subsequent disappointment'. In the absence of effective powers the schemes which were prepared endeavoured to protect rural land by zoning it for extremely low densities. By 1937 half the country was covered by such schemes and these had zoned for housing enough land to accommodate 350 million people. So much for realism! At this time there were more than 1,400 local planning authorities, half of them with populations of less than 10,000. Although there were some advisory 'Regional Plans' prepared by consultants for some joint planning committees of groups of local authorities, there was no provision for planning on a truly regional, still less on a national, scale.

Concerning these local and so-called regional plans Professor W. Ashworth has written: 'few people not professionally concerned took any interest in regional planning reports and of the small number who came across them, a substantial proportion only saw evidence that planners were spending money on the discovery of information of which they made no effective use' (Ashworth 1954).

The thirties was a period of *laissez-faire* and of great change in industrial location, hastened by the spreading of electric power which emancipated industry from its old bondage to the coalfields and by improved public and private transport which facilitated the movement of population. It was also marked by severe industrial depression in some declining areas. Neither the planning profession nor other professions had done much to identify and analyse the problems, nor was central government geared to face the realities of the situation. But there were some signs of change and of the recognition of the need for co-operation. Patrick Abercrombie, Professor of Town Planning at London and a giant among the consultants who prepared local or joint planning schemes, had as early as 1933 sought assistance from the Land Utilisation Survey. Two years later Sir Malcolm Stewart, Commissioner for the Special Areas, asked the Survey to accelerate the preparation of its maps for the Special Areas and to place them at his disposal. His second report, in 1936, paid tribute to the help he received and to the value of the maps.

In 1935 the Architectural Association, which for nearly half a century had operated a school for training architects, opened an evening School of Planning and Research for National Development. Its object was to 'widen the field of its teaching on a national scale', in the belief, as the prospectus stated, that 'the Planner capable of coping with national problems will, before long, be called into existence by the needs of the times'. It bravely expressed the hope that the School 'will succeed in welding the work of the Engineer, the Surveyor, the Architect, and the Local Government Official together with that of the Economist, the Sociologist and the Politician into that of the Planner'. (The geographer, it should be noted, was ignored.)

About fifty lecturers, many of them very distinguished, were recruited to give short courses of lectures. Perhaps there were too many teachers, but at least future planners were introduced to the importance of many problems and ideas. I myself was engaged to lecture on 'Economics as related to planning', though what I gave could better be described as aspects of geography related to planning.

Another interesting sign of concern for wider issues came in 1936 when the Town Planning Institute appointed a committee under the chairmanship of the Rt Hon. Sir Leslie Scott to prepare a report on National Survey and National Planning. Issued in 1938, this made wide-ranging recommendations and undoubtedly ensured the choice, three years later, of Scott as Chairman of the Government's Committee on Land Utilisation in Rural Areas.

A major factor in developing a public awareness of the need for planning at a national level for the control of industrial development was the realization of the severe depression, amounting to virtual dereliction, of several of the peripheral industrial areas (Willatts 1971). This had led the Government, in 1933, to appoint Sir Malcolm Stewart as a Commissioner for these areas, able to use some limited government funds to stimulate their economic recovery. The new Trading Estates at Team Valley in County Durham and at Treforest in South Wales were notable results of this new policy.

Concern for the depressed areas caused the Commissioner to sponsor special surveys and these provided an opportunity for geographers to deploy their professional skills. A good example was the 'Survey of Industrial Facilities of the North-East Coast' carried out in Newcastle under the direction of G. H. J. Daysh, Head of the Geography Department, and published by the North-East Development Board in 1936. The report of the Commissioner for that year stated that this survey by geographers was 'very favourably received and widely commended as representing a notable advance on any previous report of its kind'. Further similar surveys by the same team followed: West Cumberland in 1938 and an up-dating of Daysh's work on the North-East by A. A. L. Caesar, published in 1942.

In 1936 Sir Malcolm made the very significant suggestion that the uncontrolled growth of the metropolis was dangerous and that further factory development there should be restrained (Cmd. 5303). Indeed, there was growing public malaise at the great disparity between the economic welfare of many of the older industrial areas and that of the newer areas. At the same time there was an increasing apprehension concerning our greatest conurbation's vulnerability to hostile aircraft based in Hitler's belligerent Germany. These considerations led in 1937 to the appointment of a Royal Commission on the Distribution of the Industrial Population under the chairmanship of Sir Anderson Montague Barlow (Cmd. 6153: 1940). Its report, completed shortly after the outbreak of war, but withheld from publication until 1940, constituted the first of three great foundation stones of modern Town and Country Planning, for it proposed the decentralization of industry from congested areas and the establishment of a central national authority to provide positive direction.

The response of professional geographers, still very few in numbers, to the opportunity to present evidence to the Royal Commission was not uniformly enthusiastic. The Royal Geographical Society was invited to submit evidence and on its behalf Dudley Stamp and Eva Taylor convened a meeting of representatives of university geography departments to consider the preparation of a memorandum of evidence. I was present at that meeting, held in the boardroom of the London School of Economics. There was less than general willingness to assist. It was proposed that Stamp should act as chairman of a panel of contributors but he declined because he was committed to a five months' tour of the Far East. It was thus left to Eva Taylor, with material aid from the Director-General of the Ordnance Survey and the zeal of a small band of helpers, to prepare the submission. It consisted of a written memorandum and a portfolio of forty-nine original maps of England and Wales on the scale of 1:1,000,000 (Royal Geographical Society 1938). Although the daily minutes of oral evidence were published, war-time exigencies prevented the publication of all the written evidence given to the Barlow Commission (as it was popularly called) but the published list of submissions and the minutes of evidence, and discussion, bear witness to the wide extent of interest and thought which was generated. In retrospect, by modern standards, we may think some of the evidence was simple or commonplace, but at the time it did not seem so, either to the presenters or to the commission. I know, for I helped to draft some of it and I appeared as a witness and listened to many of the discussions.

But it is significant that at this period geographers were not suggesting that they should have a share in the process of resolving the problems which confronted the nation. In concluding a presentation to the Royal Geographical Society of her draft of evidence to the Barlow Commission Eva Taylor said, 'a study of such national maps should form a useful, if not indeed an essential preliminary to any national planning that may be under consideration'. Of that planning she had said: 'The fundamental question that must be decided is whether industry is to be forced or cajoled back into the old pattern, or whether the industrial population is to be assisted to adjust itself to the new. With that problem, however, we as geographers have nothing to do' (Taylor, 1938). Happily, in the course of the next few exciting years geographers justifiably ceased to have such reticence.

My personal concern with the Commission was, together with Stamp, the presentation of written and oral evidence based on the work of the Land Utilisation Survey. Barlow himself was impressed by the importance of the relationship between the location of industry and the availability of suitable land for industry and for agriculture and wanted a map showing the fertility of the soil. Finding that no official work had been done on this he asked us if we could prepare a Land Fertility Map. We responded with a 'Tentative Land Fertility Map of England and Wales' and a brief accompanying memorandum. Barlow was sufficiently impressed to encourage us, with a grant, to develop the work and to produce a fuller map. Thus there followed three years of intensive investigation resulting in a ten-fold classification (agreed with the Soil Survey) being employed to prepare a national map of Land Classification on the scale of ten miles to an inch (Ordnance Survey 1944; Stamp 1962).

It was not without significance that, meanwhile, the Geographical Association had elected Patrick Abercrombie (a member of the Barlow Commission) as its President for the year 1937. His stimulating presidential address (1938) was the central part of a symposium in which S. W. Wooldridge (1938) spoke on physical factors in town and rural planning and I on land use as a basis for planning (1938). In my paper I stressed the gravity of the loss of good agricultural land to uncontrolled urban development and urged that the use of rich farmland for industrial and similar purposes should require to be justified in terms of national necessity. But at that symposium we were all preaching to the converted rather than to the heathen, and our papers were not published in a planning journal but in *Geography*.

The work of the Barlow Commission stimulated widespread thought about a major national problem and thus there began a wider, wiser, and more fruitful debate on fundamental planning matters. Among geographers Stamp wrote and lectured on planning the land and Eva Taylor, appointed by the British Association for the Advancement of Science as Chairman of a Committee to prepare a preliminary plan for a National Atlas, wrote articles and gave stimulating lectures.

Soon Britain was plunged into war, at first 'phoney' but by 1940 with grim evidence that Hitler's bombers could reach not only the metropolis but all other major centres. But out of the blitz, modern planning was born. To satisfy the public fury at the destruction of our cities in the winter of 1940–1 the Minister of Works, Lord Reith, was given the responsibility for work of reconstruction and the words 'and planning' were soon added to the title of his department. Having been able to consider the recommendations of the Barlow Commission, he announced in February 1941 that his work would proceed on the assumptions that the principle of planning would be accepted as national policy and that some central planning authority would be required and that this authority would proceed on a positive policy for such matters as agriculture, industrial development and transport.

He promptly appointed an expert committee, under Lord Justice Uthwatt, to consider the vital question which bedevilled any attempt by planners to control development: that of compensation and betterment. If compensation were paid for refusal to allow development of a piece of land, the actual development took place elsewhere, but the enhanced value of the developed land was not recoverable. This committee recommended that development values on land outside built-up areas be invested in the State, on fair compensation; that interim development control be exercised on all land not covered by operative planning schemes; and that a central planning authority be established (Cmd. 6386, 1942b). Here was the second great foundation stone.

The third was the report of Lord Justice Scott's Committee which Reith, in consultation with the Minister of Agriculture, had appointed in 1941, with Dudley Stamp as its Vice-Chairman, to consider the conditions which should govern building and other constructional work in rural areas. It reported in the following year, making a wide range of recommendations, including the establishment of a central planning authority and compulsory local planning by larger units than the 1,441 local authorities who could, but were not actually required, to prepare plans (Cmd. 6378, 1942a).

Reith had previously appointed a panel of independent experts to advise him and he established within his ministry a small 'Reconstruction Group', soon to become the Planning Department of the Ministry of Works and Planning. It included a number of well-known architect–planners such as Professor (later Lord) Holford, John Dower, and Thomas Sharp, as well as George Pepler, the Chief Planning Inspector of the Ministry of Health. Two members of the Advisory Panel, Dudley Stamp and Eva Taylor, lost little time in recommending Reith to initiate a loose-leaf National Atlas, suggesting that 'the first fascicule of maps should be those of importance to planning and that the opinions of planners should guide in the selection'.

The Treasury gave approval (unenthusiastically) and the Ordnance Survey gave full co-operation, agreeing to print and publish the maps on the scale of ten miles to an inch. I was asked to take charge of the work of a research maps office, to the staff of which I was able to recruit ten enthusiastic young geography graduates. Within two years the first ten maps were in print or printing and a second instalment in progress. Thus in 1941 I became (I believe) the first professional geographer to serve as such in central government, albeit with temporary status. The nature of my work was unique for, apart from the Geological Survey, central government had no experience in compiling and producing thematic maps. I was not only responsible for preparing maps for planning purposes but for using them as research tools in applying geographical techniques to many problems of central government. The work involved the development of a major map library, most of which consisted of large-scale manuscript records.

Our enthusiasm for our work increased early in 1943 when we became the Ministry of Town and Country Planning, an entirely new government department, whose Minister was charged 'with the duty of securing consistency and continuity in the framing and execution of a national policy with respect to the use and development of land throughout England and Wales'. The staff was increased. S. H. Beaver came, to be responsible for research work on minerals. Ten regional offices were established with senior planning and research staff. Geographers were prominent among the latter and included G. H. J. Daysh, K. C. Edwards, A. A. L. Caesar, D. Trevor Williams and G. E. Hutchings. Their assistants included many younger geographers, one of whom, Mary Burns, subsequently became the County Planning Officer of Staffordshire. The work of these teams included the preparation of regional surveys for planning, termed 'planning summaries', involving the presentation, analysis and diagnosis of problems of industrial, economic and social conditions over large areas, transcending county boundaries.

Government departments were not alone in employing geographers to assist with the identification and study of problems of physical planning and propounding solutions. Several independent regional bodies did so, notably the West Midland Group on Post-War Reconstruction and Planning. This body was formed early in 1941 under the chairmanship of Raymond Priestley, Vice-Chancellor of the University of Birmingham (and subsequently President of the Royal Geographical Society), as a sequel to the Barlow report, 'to undertake a survey of the facts upon which any sound system of town and country planning in the Region must be based'.

Its reports set a very high standard of inter-disciplinary team work. Its first publication, in 1946, *English County: a planning survey of Herefordshire*, was very much the work of two members of the Department of Geography in the University of Birmingham, K. M. Buchanan and A. W. McPherson. In the following year appeared *Land Classification in the West Midland Region*, in the preparation of which the same two geographers worked with a group of soil surveyors. Both were again concerned with the Group's largest report: *Conurbation: a Survey of Birmingham and the Black Country*, published in 1948 and including an appendix by M. J. Wise.

Meanwhile, in the new Ministry, there was immense activity, frequently in matters quite new to central government. In the course of the five years 1943–7 five Acts were passed, establishing a new concept and machinery of planning. We had to consider the numerous recommendations of the three major reports already mentioned and those of scores of others, ranging from the seventy or so reports, surveys and outline plans for various regions and towns, to those of committees appointed by the Minister to advise on such diverse matters as nature conservation, access to the countryside, national parks, New Towns and the restoration of ironstone workings as well as considering the reports, with trenchantly annotated maps, on the survey of the coastline which J. A. Steers had been commissioned to undertake (Steers 1944; 1946). We worked on long-term problems ranging from the preparation of a system of Development Plans and their related surveys to those of the selection and delimitation of National Parks. We wrestled with the control of mineral workings which involved the issue of more than a score of Regulations, Orders and Circulars during the five years from 1945 onwards (Beaver 1944).

The choice of sites for New Towns was always fraught with problems, and there was considerable discussion about the right size for these towns, whether they should expand around an existing nucleus or be 'greenfields' developments, and about how to ensure that employment within them kept pace with the growth of their population. There were many interesting short-term projects, such as assisting in the selection of training areas for the armed forces, a task facilitated by the application of refined sieve-map techniques. And all the while there were the immediate and urgent problems of redevelopment, of rebuilding war-torn cities and of housing the additional one and a half million people by which the population of England and Wales had increased during the war years. Inevitably there were clashes between immediate expediency and long-term interests, sometimes enhanced by final responsibility resting elsewhere – the Board of Trade, for example, was responsible by the Distribution of Industry Act of 1945 for securing a proper balance of industry throughout the country and until 1951 housing was the responsibility of the Ministry of Health (Minister of Local Government and Planning 1951).

A problem which involved more work, and which spread over a longer period, than most others was that concerning the working of sand and gravel. At the end of the First World War the national production of gravel, principally for road metal, was less than two million tons per year. But the increasing use of concrete resulted in a tenfold increase during the next two decades with a further rise in production during the Second World War, notably to service the construction of airfields. In terms of annual mineral output it was then second only to coal. But its production was subject to no planning control. Much of it came from pits excavated in river valleys where, for example (as I demonstrated in a pilot survey in 1943) the orderly development of one growing urban district in Middlesex was largely frustrated by its being encircled by more than a dozen large unfilled lagoons.

Particularly because deposits were generally shallow, gravel extraction was making a much greater demand on land than was any other mineral, but both the public and planners generally failed to appreciate that it could only be worked where it existed. Abercrombie's 'Greater London Plan, 1944' failed to appreciate that reserves of gravel were not abundant in the metropolitan area, the growth of which had for long been steadily occluding its own building material.

Before the war it would have been unthinkable for representatives of the gravel mining industry to discuss its attendant problems with representatives of central and local government. But we were realizing that it was important to identify the reserves of the mineral in order to avoid their sterilization and to diminish conflict with agricultural interests arising from its working, particularly where gravel underlay the richest soils. It was also essential to study the consequences of its low intrinsic value and its constantly changing destination rendering it more sensitive to transport charges than any other mineral, as well as giving consideration to reducing, for example by infilling and restoration, the widespread public dislike of a generally unsightly and destructive industry.

So, within a year of the end of the Second World War and before the passing of the Town and Country Planning Act of 1947, the Minister appointed an Advisory Committee on Sand and Gravel, charged with making recommendations on future policy for the control of the extraction of this mineral. Its chairman was a distinguished engineer, Major (later Sir) Arnold H. S. Waters, V.C., D.S.O., M.C., and in addition to representatives of the gravel industry and of central and local government, two geographers, S. H. Beaver and Professor S. W. Wooldridge, were appointed as members. As the work of the committee proceeded it quickly became apparent that they made a disproportionately large input to its work, which was intense and required very extensive support from the maps office (Wooldridge and Beaver 1950).

Within eighteen months its first report was published (Ministry of Town and Country Planning 1948). The first part surveyed the general problem and dealt with such matters as the methods by which worked-out land could be reconditioned for beneficial use, the vital questions of reserves and of probable demand and of the minimization of damage to amenity and agriculture by reconditioning excavated land for planned future use. The second part was concerned with the problems of the Greater London area and included maps of the various gravel fields, showing areas proposed for gravel mining and those for reservation to agriculture.

In the ensuing six years the committee issued a further sixteen regional reports containing detailed recommendations for the solution of the general questions in each of the country's main production areas. Although its estimates of future production proved to be far too low and it failed to foresee the rapid growth in the adoption of disused wet pits for recreation of many kinds, its work was an outstanding example of a constructive survey and analysis whose adoption resulted in the development of a major industry prepared to accept planning conditions imposed in the widest public interest.

Planning developed so much during the post-war period – a direct outcome of the work done by geographers and others in the years between the wars – that in order to round off the story with which this essay is concerned it is necessary to refer, albeit briefly, to certain developments that took place in the decade after 1945 and even later, if the role of geographers is to be properly assessed. Throughout the early years of the new Ministry we had to discover how to organize our work, not only as between professional and administrative staff, but between various classes of professional staff. The new planning which was evolving required a very different approach from that which had characterized the old system. The new required team work, with a band of colleagues trained in different disciplines, each one of whom was able to contribute the appropriate wisdom of specialists whose fields of expertise overlapped his own. Clearly, the geographer's training and skills could be invaluable in the system. But this was not immediately apparent to everyone and I recall that we geographers were frequently confronted by older planners of pre-war experience who felt that our responsibility should be confined to the preparation of a survey and thereafter – and sometimes without waiting for it! – it was for the planners (that is, holders of the Diploma of the Town Planning Institute) alone to propound solutions. Their professional training for the new planning had been inadequate. It included little or nothing in the way of studies of wider environmental conditions or of social and economic matters. Many were architects and, after her earlier contacts with some of them, Eva Taylor had characteristically declared: 'scratch a planner and you find an out-of-work architect'.

By the end of the Second World War geographers engaged in planning were fully convinced that the subject needed their expertise. From practical experience they knew that it was necessary to revise drastically the modest pre-war opinion of Professor Taylor that geographers should provide surveys and leave the solution of problems to others. They were now poised to become a vital force in planning. Indeed, the most distinguished planner of the period, Professor Lord Holford (with whom some of us had worked in the Ministry) wrote in 1950 what he had said previously, that 'probably the most fundamental approach to the problems of town and country planning is that of the geographer' (Holford 1950). Lord Justice Scott, who wrote extensively on planning, had earlier declared that 'town planning is the art of which geography is the science'. But the Town Planning Institute was so slow to recognize the need for change that in 1948 the Ministry appointed a committee, under Sir George Schuster, to prepare a report on the qualifications of planners (Ministry of Town and Country Planning 1950). This, published in 1950, justified the claims of geographers, for it firmly stated that planning had become 'primarily a social and economic activity limited but not determined by the technical possibilities of design'. The Town Planning Institute responded by radically changing its examination and introducing a paper on elements of applied geology and economic geography.

Thereafter tension between differently trained members of planning teams diminished and geographers became increasingly prominent in the profession. Only sixteen years after the war a geographer (J. R. James) became the Ministry's Chief Planner and on the retirement of his successor another geographer was given the post. Meanwhile, and less than twenty years after the issue of the Schuster report, of those qualifying for membership of the Town Planning Institute no less than forty per cent had first graduated in geography. Most of them were employed by counties and county boroughs which, under the 1947 Town and Country Planning Act, became the local planning authorities. The Schuster report had said of these that their 'primary aim must be to ensure that the utilisation of land is so handled as to provide the best environment for living', because, as it also said, 'the nation ... has set itself the task of consciously regulating the setting of its social and economic life'.

From the experience of early efforts at planning during the 1918–45 period it was possible for me to identify eight principles of land-use planning when I was asked to address the inaugural meeting of the British Society of Soil Science in 1947, at the time when Parliament was enacting the Town and Country Planning Acts (Willatts 1951). These reflect the ideas of the forties and are largely the result of the work of geographers, and form the basis of modern planning practice during the four decades since the end of the Second World War. The first was the avoidance, wherever possible, of the destruction of good agricultural land. In steering his 1947 legislation through the House of Commons, Lewis Silkin declared that 'one of the main purposes of planning is to ensure that agricultural land is preserved as far as possible'. Of course, it would not always be possible, for there would be various cases of irresistible demand to satisfy uses other than for food production. A primary guide to land quality had been provided by the map of Land Classification prepared by the Land Utilisation Survey and published in the ten-mile series. Thus it was possible to ensure that disagreement about the future use of any agricultural land would not centre on its quality but on the need for its use for other purposes. It is also significant that concern for this principle ensured that the New Towns, built following the New Towns Act of 1946, used less land per head for all purposes than did existing towns. It took another quarter of a century for the land classification work of the Land Utilisation Survey to be superseded by more detailed official work (Ministry of Agriculture, Fisheries and Food 1967–75).

The second principle was that the physical development of towns should be reasonably controlled and uncertainty about the future of undeveloped land should be removed. There had been justifiable criticism of the way in which land on the fringes of towns, often advertised as 'ripe for development', ceased to be effectively farmed because of the possibility of its sale for building. The 1947 Act sought to end this by requiring local planning authorities to designate the land which would be needed for justifiable development in the next ten years. To quote the Minister again 'if land is not designated at least there is some guarantee that it will not be required in the course of the next ten years'. A related principle is that due regard should be had to the need to avoid severing farm units.

Third, the size of settlements should take account of various factors, including the need to avoid coalescence of urban nuclei and to preserve peripheral belts of open country. A few years later the ministerial enunciation of a Green Belt policy sharpened the application of this principle.

A fourth principle was the recognition of the intimate relationship of most towns with their neighbouring rural settlements, provision for which fuller economic, social and cultural life must be made in the urban centres, which themselves should constitute the hubs of rural life. Before the war it would have been hard to find, in the work of planners or of geographers, much appreciation of the complex regional inter-relationships of town and country, but in the forties and fifties this was rectified, notably by work carried out in the Ministry's maps office, under the direction of F. H. W. Green (Green 1950; Ordnance Survey 1955).

Fifth and very important, was the principle of avoiding the sterilization of valuable mineral resources, an objective which had been generally disregarded. Vast deposits of our chief minerals, especially coal and gravel, had been 'sterilized' by urban development. The establishment in 1946 of the National Coal Board facilitated the co-ordination of plans for physical development on the surface with those for effective winning of coal beneath. In the same year the Minister of Town and Country Planning appointed the Advisory Committee on Sand and Gravel (mentioned above). Other committees and conferences dealt with various other minerals (Beaver 1949).

Related to this was a sixth principle, that of restoring to beneficial use land from which minerals have been worked. Before the war this had received but scant attention; indeed in 1939 a government-appointed committee had concluded that, because of the excessive cost involved, it would not be possible to restore to agriculture the 150 or so square miles of England from which iron ore was expected to be won by opencast working (Ministry of Health 1939). A new climate of public opinion, appropriate legislation, and the rapid war-time development of new machinery changed all that, as also with opencast coal mining. The change in respect of gravel working, which expanded most dramatically in the new age of concrete, was striking. After the 1947 Act planning consents became conditional on site restoration. This was not always restoration to agriculture but in river valleys where excavation resulted in lagoons, pits have been adapted for many kinds of water recreation, for which there continues to be an unsatisfied demand.

The seventh principle was that planners should always have regard for such matters as relief, structure, soil and drainage and related factors, such as microclimate. I recall many attempts to convince planners of the significance of such basic considerations. The work of educating them, particularly on climatic matters, was far from easy but it became less difficult with the steady increase of relevant studies by independent geographers.

The eighth principle was that of multiple use of land, ensuring that much of the country serves more than one purpose concurrently. Most military training areas can be farmed, and both they and forestry areas can be used for recreation, as can water-gathering grounds and reservoirs. The erosion of opposition to this principle (for example, from water authorities) and its steadily increasing application would in itself constitute a fascinating study and is one of the important achievements of land-use planning.

Looking back now to the first decade of modern planning I realize how easy it would be to criticize much that was done and to point to what should have been done but was not done. Defects always become clearer with hindsight. But it cannot be denied that within a relatively short time a great deal had been achieved and planning had advanced from a system of preventing certain obvious evils to one of promoting the desirable environment for living. This major change had provided geographers with opportunities to apply their professionalism in new ways for the advancement of the common good.

REFERENCES

P. Abercrombie (1938), 'Geography, the basis of planning', Geography, 23, 1–8.
W. Ashworth (1954), 'The genesis of modern British town planning', Routledge and Kegan Paul.
S. H. Beaver (1944), 'Minerals and planning', Geographical Journal, 104, 166–93.
    (1949), 'Surface mineral working in relation to planning', Report of the Town and Country Planning Summer School, The Town Planning Institute, 105–30.
H. C. Darby (1983), 'Academic geography in Britain, 1918–46', Transactions of the Institute of British Geographers, N.S. 8, 21.
F. H. W. Green (1950), 'Urban hinterlands in England and Wales: an analysis of bus services', Geographical Journal, 116, 64–88.

W. G. Holford (1950), *Town and Country Planning Textbook*, Association for Planning and Regional Reconstruction, London.

J. R. James, S. F. Scott and E. C. Willatts (1961), 'Land use and the changing power industry in England and Wales', *Geographical Journal*, 127, 286–309.

Ministry of Agriculture, Fisheries and Food (1967–75), *Agricultural Land Classification Maps of England and Wales:* Provisional maps on the scale of one inch to a mile prepared by the Agricultural Development and Advisory Service of the Ministry of Agriculture, Fisheries and Food with the assistance of the Soil Survey of England and Wales.

Ministry of Health (1939), *Report of the Committee on the reconstruction of land affected by iron ore working*, H.M.S.O.

Minister of Local Government and Planning (1951), *Town and Country Planning, 1943–51*, Cmd. 8204, H.M.S.O.

Ministry of Town and Country Planning (1948), *Report of the Advisory Committee on Sand and Gravel, Parts 1 and 2*, H.M.S.O.

(1950), *Report of the Committee on qualifications of planners*, Cmd. 8059, H.M.S.O.

Ministry of Works and Planning (1942a), *Report of the Committee on land utilisation in rural areas*, Cmd. 6378, H.M.S.O.

(1942b), *Final Report of the expert committee on compensation and betterment*, Cmd. 6386, H.M.S.O.

Ordnance Survey (1944), *Great Britain: Land Classification, 1/625,000 or about 10 miles to one inch* (2 sheets). Also related explanatory Text No. 1.

(1955), *Great Britain: Local Accessibility: the hinterlands of towns and other centres as determined by an analysis of bus services* (2 sheets), *1/625,000*. Also related Explanatory Text No. 6.

Report of the Commissioner for the Special Areas (England and Wales) (1936), Cmd. 5303, H.M.S.O.

Royal Commission on *The Distribution of the Industrial Population (1940), Report*, Cmd. 6153, H.M.S.O.

Royal Geographical Society (1938), 'Memorandum on the geographical factors relevant to the location of industry', *Geographical Journal*, 92, 499–526.

L. D. Stamp (1962), *The Land of Britain: its use and misuse*, 3rd edn, Chapter XVII.

J. A. Steers (1944 and 1946), 'Coastal preservation and planning', *Geographical Journal*, 104, 7–27, and 107, 57–60.

E. G. R. Taylor (1938), 'Discussion on the geographical distribution of industry', *Geographical Journal*, 92, 22–39.

E. C. Willatts (1938), 'Present land use as a basis for planning', *Geography*, 23, 94–105.

(1951), 'Some principles of land use planning', *London Essays in Geography: Rodwell Jones Memorial volume:* London School of Economics.

(1952), 'L'état actuel de la planification régionale en Grand-Bretagne et la contribution des geographes', in *L' aménagement de L'espace: planification regionel et geographie*, ed. J. Gottman, Cahiers de la Foundation Nationale des Sciences Politiques, 103–33.

(1971), 'Planning and geography in the last three decades', *Geographical Journal*, 137, 311–38.

S. W. Wooldridge (1938), 'Town and rural planning: the physical factors in the problem', *Geography*, 23, 2: 90–3.

S. W. Wooldridge and S. H. Beaver (1950), 'The working of sand and gravel in Britain: a problem in land use', *Geographical Journal*, 115, 1: 42–57.

9   On the writing of historical geography, 1918–1945

H. C. DARBY*

I am not sure when the term 'historical geography' was first used. One early example comes from 1834 when it appeared in the phrase 'historical or political geography' in the article on geography in the seventh edition of the *Encyclopaedia Britannica*. Another early example is dated 1846 when it entered into the title of Karl von Spruner's pioneer historical atlas. In this context it implied concern with changes in political boundaries and with the varying extent of states and provinces; and this usage has continued among some people up to the present day. It also formed part of the titles of a number of books in the 1840s which were very largely historical topographies.[1]

Before the end of the century the term was also used to indicate concern with the influence of geography upon history. That it became increasingly frequent may be gathered from the fact that in 1873 H. F. Tozer (himself a classical geographer) could say that A. P. Stanley had 'done more than any living man to promote the intelligent study of historical geography'; his most notable contribution was a volume on *Sinai and Palestine* (1856), which sought to trace the relations between the geography of the area and the history of its people.[2] Stanley acknowledged the help of 'Mr Grove, of Sydenham' in Kent, who later contributed many geographical articles to William Smith's *Dictionary of the Bible* (1860–5). This was none other than George Grove, later famous for his *Dictionary of Music and Musicians* (1878–97). When Grove received an honorary degree at

* Henry Clifford Darby, C.B.E., F.B.A. (b. 7 February 1909) graduated in geography at Cambridge in 1928. From 1932 until 1945 he was a University Lecturer in Geography and a Fellow of King's College, Cambridge. In 1945 he was appointed John Rankin Professor of Geography at Liverpool, a post he held until 1949 when he moved to University College London. In 1966 he returned to Cambridge on his appointment to the Chair of Geography. He retired in 1976 when he became Professor Emeritus. He is an Honorary Fellow of St Catharine's College and of King's College.

Durham in 1875, reference was made to 'his writings in that branch of Biblical learning which related to historical geography'; and the citation went on to describe these writings as 'characterised by that combination of physical and historical enquiry which had marked the gifted German geographer Karl Ritter'. Then in 1894 appeared what may possibly be the most remarkable book ever to include the term in its title – George Adam Smith's *The Historical Geography of the Holy Land*. Its aim was 'to discover from "the lie of the land" why the history took certain lines'. It reached its twenty-fifth edition in 1931, and was reprinted even as late as 1966 and 1973.[3]

After 1900 came other books dealing with the influence of geographical conditions upon historical events. Such was James Fairgrieve's *Geography and World Power* which appeared in 1915, and which was reprinted a number of times in the 1920s and later.[4] Among other books with a similar theme were Rodwell Jones's 'historical geography' of *North America* (1924) and Marion Newbigin's 'human and historical geography' of *The Mediterranean Lands* (1924). More unusual was Vaughan Cornish's *The Great Capitals* (1923) with the subtitle 'historical geography'.[5] Other notions were also finding shelter under the umbrella of the term – the history of exploration, or of cartography, or of geographical study itself, but the two main ideas current in the early decades of the century were changing frontiers and geographical influences.

This was the kind of historical geography I encountered when I went up to Cambridge in 1925 to read geography. The University Regulations of the time defined it as follows:

> *Historical and Political Geography.* The geographical conditions affecting the historical and political development of States; movements of population and centres of influence; frontiers; colonial expansion; political subdivisions for administrative purposes.

I attended the classes of Mr B. L. Manning of Jesus College who had become a lecturer in geography in 1921. That he was a young historian was nothing unusual. Geographers such as P. M. Roxby, J. N. L. Baker, and W. G. East started with degrees in history. Manning's heart, however, was in ecclesiastical history to which he later made distinguished contributions, and he resigned his post in 1930 to become a lecturer in the History Faculty. In the meantime, his lectures to us were of the kind that has been called 'an historian's historical geography'.[6] We began with Ancient Greece, then went on to the Roman Empire and so to the Byzantine Empire. We continued with the territorial evolution of France, with feudal Germany, with the rise of Russia, with the many states of Italy before

1870, with Spain and Portugal, and with the development of Muscovy into Russia. We then embarked upon the expansion of Europe overseas. Our reading at this time included H. B. George's *The Relations of History and Geography* (then in its fifth edition) and E. A. Freeman's *The Historical Geography of Europe* (then in its third edition) which 'traced out the extent of different states at different periods'. I recall spending hours disentangling the various connotations of the name Burgundy – the kingdoms, the duchies, the county, and the imperial *Kreis* or 'circle' of later times; little wonder that Freeman wrote: 'no name in geography has so often shifted its place and meaning'. An essential part of our equipment was an historical atlas, and so I acquired W. R. Shepherd's *Historical Atlas* (also in its third edition). One book that made a great impact on us was Lucien Febvre's *A Geographical Introduction to History* which appeared in English in 1925. Other books which we read were straight histories, and I do not regret spending a fair amount of time reading J. B. Bury's fascinating *History of Greece* which had been reprinted for the sixteenth time in 1924.[7] (I mention the editions to show that these books had great staying-power.) Looking back, I realize how well Manning's lectures were delivered, and how much I gained from them. As a contribution to a general education they were superb.

When I began to work for a Ph.D. in 1928, it was natural that my interests should reflect the tradition in which I had been taught. My thesis was entitled 'The role of the Fenland in English history' – the part it played as a barrier between East Anglia and Mercia, and as a camp of refuge in times of rebellion. My examiners in 1931 were a geographer and an historian, P. M. Roxby and J. H. Clapham; and they seemed to see nothing wrong with the method; some parts of the thesis appeared later in historical publications.[8] This was the first Ph.D. in geography awarded at Cambridge. Within a few years such a thesis on geographical history would certainly not be presented from the Department of Geography.

The thesis did, however, contain a chapter concerned with 'reconstructing the past environment of the Fenland', and this appeared in the following year in the *Geographical Journal*.[9] Moreover, the introduction said that 'the scope and content' of historical geography were 'far from agreed upon', and it expressed uneasiness with the theme of geographical influences. The next few years were to see these doubts resolved.

Historical geography and the 'New Geography'

In the meantime much had been happening in the outer world to affect the place of geography as an academic discipline. There was, among other things, the delayed impact of Darwinism and the continuing influence of industrial change, upon social as well as upon scientific thought. Attention now turned to the environment in relation to man. Geographical teaching had started in Oxford and Cambridge as early as 1877–8, but for many years this was supplementary to that in other subjects; honours schools still lay in the future. Chairs were established at University College London in 1903, and at Liverpool and Aberystwyth in 1917. Other Universities soon introduced geography into their programmes. The phrase 'New Geography' seems to have been used by H. J. Mackinder as early as 1886, but the idea it implied did not come to full fruition until after 1918.[10] In retrospect, we can see that the years following the end of the First World War saw the take-off of academic geography from its early beginnings into fairly sustained growth. This brought with it a fresh consideration of the relations between geography and history. Whereas, hitherto, some historians had believed in the relevance of geography to history, now some of the increasing number of geographers in Britain began to reverse the thinking and to consider the relevance of history to geography.

Ideas often have their early anticipations. In 1863, at King's College London, William Hughes had written a book on *The Geography of British History* with a sub-title which ran: 'a geographical description of the British islands at successive periods from the earliest times to the present day'. Five of its chapters attempted cross-sections – Roman, Saxon, Norman, Tudor and the 'present-day' of the 1860s. More in the mainstream of the subject was the course of six lectures – five given by historians – organized by Herbertson at Oxford in 1906, each devoted to the geography of a period. One of these, on Roman Britain, was by F. J. Haverfield who incorporated the substance of it into his Ford Lectures in 1907 which were published in 1924. J. N. L. Baker, who recorded these facts, concluded: 'Here we have historical geography in its modern sense.'[11] We have it on Mackinder's authority that, 'on the historical side', Herbertson's equipment was 'weak'.[12] That may have been so as far as technical scholarship was concerned, but at any rate in his copy of Freeman's 'Historical geography', Herbertson could write that 'Historical geography describes and interprets human distributions at any past period, and the successive changes of human distributions, economic, political, and racial in the widest sense, within a defined area throughout historical time'.[13] And in his last paper, published in 1915, he found 'the present conditioned by the past' and thought it necessary for a geographer to understand 'the phases of development' a region had passed through.[14]

Other geographers were also on the scent. In 1907, J. F. Unstead described historical geography as cutting 'historical sections through time' in the same way that 'ordinary geography cuts through time at the present period'.[15] In 1922 he defined it as 'the geography of the past', but went on to speak of 'continuous development' and of 'the study of the evolution of man's environment', and of the interactions, at each stage of evolution 'between nature and man'.[16] If I read them aright, these remarks imply a duality – on the one hand, the cross-sectional description of an area; on the other, a narrative of change through time. A similar duality may be seen in the statements of other writers. Rodwell Jones spoke in 1925 of the need to study 'a few selected periods, or regions', that is to say, cross-sections or narratives.[17] Not long afterwards, Roxby was speaking not only of 'the reconstruction of the physical setting of the stage in different phases of development' but of 'the evolution of the relations of human groups to their physical environment'.[18]

Some aspects of the changing thought of these years may be illustrated from the work of H. J. Mackinder. His *Britain and the British Seas* was published as long ago as 1902 but it was reprinted in 1907 and 1915. One chapter deals with 'historical geography'. Much of this is devoted to the Anglo-Saxon and Scandinavian settlements and their influence upon the origin of the counties and dioceses of the British Isles. The chapter concludes by saying: 'The geography of Britain is in fact the intricate product of a continuous history, geological and human.' Interesting though it is, the chapter does nothing to prepare us for the human geography of the 'present-day', that is of 1902. Elsewhere in the book there is much about geological history, and there is a chapter entitled 'The physical history of Britain', but there is nothing about the history that produced, for example, the varied lay-out of field and hedgerow, the disposition of arable and grass, or the location of industry. By 1928, it was a different Mackinder who wrote: 'There is however a true historical geography. It involves what literary people call the historic present. The historical geographer seeks to restore imaginatively the dynamic system of some past moment of time.' And three years later he returned to the theme of 'the historical present'.[19]

An opportunity for the 'New Geography' of the post-1918 years to show its strength came in 1928, when the Twelfth International Geographical Congress was held in Cambridge. To mark the occasion, the University Press published *Great Britain: Essays in regional geography*. It was edited by A. G. Ogilvie, and written by twenty-six authors drawn mostly from sixteen departments of geography. The opening paragraph of the Introduction stressed the need 'to know something of the history both natural and human of the country, in order to understand and appreciate the things one sees'. One of the chapters tells us that 'historical geography' covers two distinct ideas; one a review of history from the point of view of a geographer; the other, an account of geographies of past periods. It added that the first approach was the more usual.[20] When we look at the chapters in detail, what a variety of method meets our eyes. Some chapters make hardly any reference to human history, yet we cannot believe that they objected to an historical approach on principle because they certainly made substantial excursions into geological history. Other chapters include sections with such headings as 'historical geography', 'historical and human geography', 'notes on historical geography', 'population and historical change' and 'the human evolution of the region'. Yet other chapters are without such specific historical sections but include retrospective allusions as and when they seemed appropriate. Such variety is often characteristic of a co-operative work. Even so, there is much that is random or illogical about some of the writing. The historical paragraphs in some chapters are interesting in themselves, but they are not relevant to any explanation of why a region was as it was. Furthermore, some chapters tell us a great deal about, say, prehistoric times or about the Anglo-Saxon period, but hardly anything, sometimes nothing, about the agrarian and industrial changes that resulted in 'the things one sees'. Clearly there was much diversity in the aims and methods of those who wandered about in the territory between geography and history. As for the Congress itself, the papers delivered to the section on historical geography dealt mainly with the history of cartography and discovery.[21]

Two years later, in August 1930, there took place the 'First International Congress of Historical Geography'. That was its official title – *Premier congrès international de geographie historique*. It was held in Brussels and was attended by about 200 people, mainly historians but with some geographers. The great majority of the members came from Belgium itself, but there were eleven from Britain, and these included the geographers Arthur Davies, W. G. East, G. H. T. Kimble, J. H. G. Lebon, E. G. R. Taylor, and myself; there was also Sir Charles Oman, professor of modern history at Oxford. The majority of the sixty or so papers delivered to the various sections dealt with the history of cartography and with boundaries – administrative, ecclesiastical and linguistic.

On the first day of the Congress came a very depressing paper (not delivered but later printed in the Proceedings) by Sir Charles Oman entitled 'Note on geography as applied to history in Great Britain' in which he said: 'I sincerely wish that I could give a more favourable account of the manner in which history and geography are now related to each other in this country.' And he went on to add: 'So far as I can see there is no attempt made to correlate the two branches of learning to each other.' There may have been something in what he said but, clearly, he was quite unaware of the stirrings that were being prompted by the rise of academic geography in the 1920s. On the next day, however, came a paper entitled 'Diverse conceptions of historical geography'. It was given by the Brussels archivist Charles Pergameni, and it put forward a plea for recognizing one meaning of the term as 'the human geography of the past', and it referred to the work of the German Alfred Hettner who had spoken of 'past geographies'. The 'very substantial discussion' that followed was generally in favour of this view, which stood out as a fresh approach among the older views represented by the Congress as a whole.[22] At any rate it seemed a fresh approach to one who was struggling with his doubts as he completed his Ph.D. thesis.

Cross-sections

In Britain, the advance of the 'New Geography' was producing its own ferment, and the rising generation of geographers was anxious to clarify and define its position in relation to that of other disciplines. One expression of this feeling was a 'well-attended' meeting in London in January 1932 between representatives of the Geographical and Historical Associations to discuss the question 'What is historical geography?' While recognizing that the term was currently used in a variety of ways, the geographers (but not the historians) were emphatic in believing that logically it could only mean, in the words of J. N. L. Baker, 'the reconstruction of the geographical conditions of past times'. Or, as E. G. R. Taylor put it: 'The application of the adjective "Historical" to the noun "Geography" strictly speaking merely carries the geographer's studies back into the past: his subject matter remains the same.'[23] Among those present was E. W. Gilbert, and, later in the year, he expanded his statement in a separate paper on the same question which he answered by saying: 'The real function of historical geography is to reconstruct the regional geography of the past.'[24]

Another symptom of the ferment of the time was a paper by W. G. East in 1933. This was, in a sense, 'a thinking aloud' about some of the problems presented by 'period pictures'. Should regions be adopted as a basis, and, if so, how should they be distinguished? What, and how many, 'culture periods' should be selected? Should the treatment be extended backwards into periods for which archaeological evidence alone was available? And he concluded by envisaging 'a whole series of past geographies which culminate in the present-day geography, itself destined to disappear'.[25] Some of these ideas he put into practice in *An Historical Geography of Europe* which was first published in 1935. As the preface says, the treatment could only be selective; but it marked a distinct step forward from earlier English books bearing the title 'historical geography'.

Thus it was that with the development of academic geography in Britain in the 1920s and 1930s came the rise of historical geography as a *self-conscious* discipline. We 'new geographers' realized that every past had once been a present. There was a high degree of unanimity among us. We had something of the dogmatic fervour of new converts to a faith, heightened by the fact that the position of geography as an academic discipline was not all that well-established. Being insecure, we were emphatic.

We were, moreover, dissatisfied with professing the new faith without attempting good works. When I suggested the idea of a co-operative volume on England to some colleagues in other universities, they warmly welcomed it, and we met in London to consider a scheme. Some of us, while strong in support, were doubtful about the outcome. 'Who will publish it?' asked one. 'Who will buy it in these days of depression?' asked another. (At this point may I say that the total full-time geographical teaching staff in British universities was under ninety.) And so I returned to Cambridge with an agreed plan, but in some uncertainty. By this time (it was February 1934) I had succeeded Manning in his University lecture-ship in geography, and had also become a Fellow of King's College where I got to know J. H. Clapham. He was the first professor of economic history at Cambridge, and was not only a distinguished scholar (he later became President of the British Academy), but a man of great administrative ability. As an economic historian he was interested in human geography, and was well aware of French work in the subject. Only a few years earlier he had written: 'It is much to be desired that there should be a close union between the two subjects.'[26] One evening, I told him of our meeting, our plan and our problem. His reaction was immediate: 'It is an interesting idea. Why not offer it to the Press? I will tell the Secretary about it tomorrow morning.' He could say this because he was a Syndic of the Cambridge University Press. Within a matter of days the arrangements were in train. We got to work at once, and the book was published in June 1936.

Thus it was that eleven of us joined together to produce *An Historical Geography of England before AD 1800*. Our aim was not to produce some broad general views, but geographical descriptions based as far as possible on primary sources; and our hope was to match the scholarship of contemporary historians. We were very anxious that all the contributors should be professional geographers, but we soon realized that there was no British geographer with expertise in the Scandinavian period. I therefore invited Eilert Ekwall, professor of the English language at Lund in Sweden, to contribute a chapter. He was a distinguished authority on the place-names of England, and he produced a most valuable contribution for us; but at the time, I was sorry that we had to go outside our own ranks. The ten geographers came from four universities that had done much for the subject – Oxford (Baker, Gilbert), Cambridge (Darby, Spate), London (East, E. G. R. Taylor, Wooldridge), and Aberystwyth (Bowen, Pelham, D. T. Williams, the last two of whom had moved to other universities by 1936).

That the volume ended in 1800 had no methodological significance. The date was chosen because the geography of the nineteenth century had recently been covered in certain chapters of Clapham's *An Economic History of Modern Britain*, two chapters entitled 'The face of the country', in 1820 and in 1886–7. At any rate, 1800 was a convenient date, and we seemed to have lacked either the equipment or the resolution to go beyond it. The Preface described the volume as 'in a sense, experimental', and said that, quite deliberately, no attempt was made to provide a philosophical introduction. I well remember thinking about this for months as the book passed through the proof stage, but coming to the conclusion that the time was not ripe for a methodological essay. Maybe I lacked the nerve. In the light of this hesitation it is interesting to note that when Marc Bloch reviewed the book in the French *Annales*, he began by saying: 'Our vocabulary is so imperfect that to entitle a book "An historical geography" is to risk not giving in advance a very precise idea of its content.'[27]

Some twenty years or so later, I looked back upon the enterprise in what I hope was a quite dispassionate manner.[28] By then the book had been reprinted a number of times, and may have, as the phrase goes, 'fulfilled a need'. But I had to recognize that, methodologically, the 'experiment' had been only partially successful. The Preface spoke of 'the reconstruction of past geographies' and of 'a sequence of cross-sections taken at successive periods'. The volume did contain some of these, but, on the other hand, some chapters were concerned not so much with cross-sections but with developments through time – the Anglo-Saxon settlement, the Scandinavian settlement, the draining of the Fens, the growth of London. The inclusion of these narratives can be vigorously defended, but they cannot be called cross-sections. Nor could they have been otherwise. The form they took was dictated partly by the nature of the periods they spanned, and partly by the available source material.

We certainly know now that the making of cross-sections is far more complicated than some of us realized in 1936. There can be thin cross-sections and thick cross-sections. Some cross-sections are so wafer-thin that, paradoxically, they lack an historical approach. Such instantaneous cross-sections, especially in the form of period maps, may be useful for some purpose, but clearly they do not enable us to appreciate the processes of change that result in a landscape and its geography. Other cross-sections are so thick that they partake of the nature of narratives. One possibility is to alternate descriptive cross-sections with explanatory narratives. Yet another is to make one's cross-section go backward in time instead of forward. Furthermore, one must always remember the distorting effect of hindsight upon the evaluation of contemporary conditions in past ages. All these various approaches have been discussed from time to time.[29]

Between 1936 and 1945, the cross-sectional method came to the surface in a variety of ways. One of these was W. G. East's studies of the land utilization of various counties based upon the Board of Agriculture Reports of around 1800. Some of them appeared in a paper of 1937, and others followed in various Reports of the Land Utilisation Survey and elsewhere.[30] There were also experiments in presenting the Domesday information for a number of counties; and in 1937 came a plan for a complete coverage of Domesday England, a plan that went into abeyance during the war. Then in 1939 there appeared the *Historical Geography of Southwest Lancashire before the Industrial Revolution* by F. Walker, one of Roxby's students in the Liverpool department; and in 1940 came an account of the medieval Fenland, an expanded version of my paper of 1932.[31] In the following year Bowen's study of the 'geography and history' of Wales included five brief cross-sections of 'the cultural landscape' at important periods in Welsh history.[32]

Geographical changes

An interesting feature of these developments in the 1930s is that while we believed so fervently in cross-sections, many of us were also engaged in other activities in the borderland between geography and history. We were describing processes, and writing narratives that were sometimes accused of being nothing other than economic history. In the daytime, so to speak, we practised cross-sections, and in the evening we indulged in narratives – the horizontal versus the vertical, pattern versus process. The latter was a reasonable approach if one regarded the geography of the present-day as a collection of legacies from the past. This was the idea implied in the term 'cultural landscape', and some thought of historical geography as being an integral part of cultural geography which was often equated with human geography. C. B. Fawcett had meant something similar when he wrote in 1932: 'Historical geography is essentially that part of geography in which we are studying the influence of historical events on geographical facts.'[33] One can appreciate his point without accepting his definition.

The relevance of an historical approach raised what, to some geographers, was an important question: 'How much history?' It was J. F. Unstead who in 1922 had laid down what he called 'a general principle', when he wrote: 'As Geography deals with present conditions, the past is only to be evoked when it is necessary to explain the present.'[34] He thought that 'those parts of man's history which affect the geography of today' should form an ingredient of geographical writing just as 'physiographic history' does. It followed from this, he said, that 'much of the information about earlier times is irrelevant' to the present day, and so should be excluded from a geographical account. Such irrelevant information, he added, 'might well be used as material for a study in Historical Geography with a different object in view', that is, the reconstruction of the geography of a past age. Mackinder, in 1931, was also concerned that we should not 'mix history with geography without seeing clearly what we are doing'. While admitting that 'in the geography of today are undoubtedly a number of remnants' of past geographies, he thought that this 'fact should not alter the whole perspective of the main subject'.[35]

Some geographers, however, were not deterred by the difficulty of making such a 'nicely calculated less or more' that these cautions enjoined. Many of their studies dealt with the changing character of specific countrysides.[36] An outstanding example, in 1933, was E. C. Willatts' study of changes in the south-west of the London Basin, 1840–1932. He sought 'by examining the past, to understand more fully how the present landscape has been evolved', and to do this, he compared the tithe maps of about 1840 with those of the Land Utilisation Survey in the early 1930s. Three years later came another valuable study by H. C. K. Henderson in which he was able to use the Ordnance Survey Area Books of 1875, as well as the tithe maps to show the changing agriculture of the Adur Basin, Sussex. The possibilities of another category of information – eighteenth-century estate plans – were used by Arthur Geddes (1938) and by J. H. G. Lebon (1946) in their accounts of the changing landscapes of the Lothians and of Ayrshire respectively. Among other examples there was my own account (1940) of the draining of the Fens.

Other studies of change by geographers were devoted to towns.[37] The titles of many of these included such phrases as the evolution, the origin and development, the historical geography, the rise and growth (sometimes simply the growth) of such towns as Bristol, Cambridge, Dundee, Hull, Sheffield, Whitby, as well as 'inland and seaside health resorts'; there were also studies of London and its port. They were written in a variety of contexts and to meet different occasions; but it is clear that those which used the term 'historical geography' did so to imply a narrative of change and as synonymous with growth and development.

Yet other studies were concerned with industrial changes, and were described by such words as 'development' or 'historical geography'. They dealt, for example, with the iron industry of the Forest of Dean, with the coal industry of Coalbrookdale and of Northumberland and Durham, and with the Cotswold woollen industry.[38] With these may be grouped a number of studies dealing with the movements of population within Britain by R. A. Pelham and others.[39]

Historical geography by non-geographers

While many geographers were restricting the term 'historical geography' to describe the geography of past periods, some historians were also incorporating geographical descriptions – or cross-sections – in their own studies. Just as Monsieur Jourdain in Molière's *Le bourgeois gentilhomme* discovered that for forty years he had been speaking prose without knowing it, so these historians were writing historical geography without being aware of the fact; the example we often quoted in the 1930s was the third chapter of Macaulay's *History of England* (1848) with its description of England in 1685. Their delineations may not have been exactly the reconstructions that a geographer would produce – the geographer with his preoccupation with physical circumstances and with distribution maps. But the difference was not one of principle, and frequently the work of the geographer and the historian was hardly to be distinguished, one from the other.

Historians contemporary with ourselves were hard at work on their cross-sections.[40] Clapham, as we have seen, included two accounts of what he called 'the face of the country' as part of his large-scale *Economic History of Modern Britain*. Both G. M. Trevelyan and G. D. H. Cole wrote descriptions of England in the eighteenth century, based largely upon Daniel Defoe's *Tour*. David Ogg gave a picture of 'The Land: its products and industries' in his account of seventeenth-century England. During these years A. L. Rowse was also at work, thinking 'back to what the face of the country looked like' in Elizabethan times; and S. T. Bindoff was shortly to describe how England in Tudor times 'wore an appearance very different from that which the name now conjures up to our minds'.

In addition to these surveys of England in the sixteenth, seventeenth, eighteenth and nineteenth centuries, there were others of particular districts.[41] W. H. B. Court's account (1938) of the Midland industries begins with 'a sketch of the resources, natural and above all human', and so provides an excellent cross-section of the geography of the area on the eve of its transformation into a great industrial region. Furthermore, certain chapters of J. D. Chambers's book on Nottinghamshire (1932) in the eighteenth century, based as they were upon an appreciation of regional differences within the county, go a long way towards providing a geographical survey. Or yet again, T. S. Willan's two books (1936–8) on river navigation and on coasting trade (both concerned with the seventeenth and eighteenth centuries), are studies which no historical geographer of England can afford to neglect. Different in character was William Rees's scholarly map of South Wales and the Border in the fourteenth century – on a scale of two miles to one inch, and in four sheets. He described it as 'a study in historical geography', and it appeared in 1933, nine years after his book on the same subject.

Unstead had said in 1907 that 'classical geography is a special case of historical geography'. The work of two historians of the ancient world certainly made a great impression upon us in the 1920s and 1930s.[42] One was Alfred Zimmern whose *Greek Commonwealth* had first appeared in 1911 but with third and fourth editions in 1921 and 1924. Part One of the book was entitled 'Geography' and it provided an outstanding picture of the Athenian world in the fifth century BC. Here, we recognized, was historical geography of the first quality. Then there was J. L. Myres whose interests were so geographical that we almost forgot he was a professor of ancient history. He was frequently to be seen at geographical meetings; and in 1928 he became President of Section E (Geography) of the British Association which in those pre-Institute of British Geographers' days was an important forum for geographical discussion. His address was on 'Ancient geography in modern geography', and was one of the many geographical papers that came from his pen. Among other studies of the ancient world was M. P. Charlesworth's account of the trade routes and commerce of the Roman empire (1924), a searching enquiry which ran into a second edition in 1933 when a map was added.[43]

Just as historians sometimes needed to portray the geography of a past age for their own purposes, so did archaeologists and prehistorians. Their work may be regarded as a special case of the cross-sectional method – the creation of what the Germans were calling the *Urlandschaft*. There had been a number of pioneer studies along these lines during the nineteenth century, and in 1915 came J. P. Williams-Freeman's book on the archaeology of Hampshire which included a chapter dealing with the 'natural conditions of the country' and a Quarter-Inch map showing 'the distribution of forest, scrub and treeless country' upon which archaeological distributions were plotted.[44] It was this book that inspired O. G. S. Crawford's 'attempts to restore the natural vegetation, chiefly woodland, on a geological basis'.[45]

Crawford had taken the Oxford Diploma in Geography in 1910, and, as he wrote later, he had been 'much mixed up with the early developments of human geography and the Oxford school of Professor Herbertson'.[46] His future, however, lay not with geography but with archaeology, and into his new vocation he carried his interest in 'the employment of the distribution-method'. His service during the 1914–18 War involved him in the interpretation of aerial photographs which made him realize the great possibilities of this new technique for archaeological work. After the war he became Archaeology Officer to the Ordnance Survey, and his tenure of this post (1920–45) was notable for his use of aerial photography and for the idea of 'period maps'. A total of fourteen such maps were published up to 1938, including one on Roman Britain (1924; 2nd edn 1931), two on Britain in the Dark Ages (1935, 1938), and one on seventeenth-century England (1930).[47]

The reconstruction of primitive vegetation on some of these maps was sometimes criticized, and it was omitted from the third edition of the map of Roman Britain (1951). Even so, it often helped in the interpretation of archaeological distributions over more limited areas, and it was used to great effect in Cyril Fox's study of the Cambridge region (1923), with its five Quarter-Inch maps of phases of occupation from the Neolithic to the Anglo-Saxon period. Nine years later came Fox's *The Personality of Britain* (1932) which aroused great interest. Carl Sauer, for example, thought it an 'admirable study', and he borrowed its title when he wrote 'The personality of Mexico'.[48] Taken together, the work of Crawford and of Fox is assured of a place in the pantheon of British historical geography.

Many non-geographers were producing not cross-sections but narratives of change. It is true that in 1923 the geologist R. L. Sherlock could say that while much had been written about the effect of nature on man, 'it is remarkable that the effect of Man on Nature seems to be almost entirely ignored'.[49] He was, of course, thinking mainly of man as a geological agent of denudation and deposition, and he did less than justice to the numerous studies dealing with engineering and agricultural topics, with, for example, draining and irrigation and with deforestation and enclosure, even though his bibliography contains many references to these.

By 1945 Sherlock's statement could not have been made because of the increasing number of studies of the way in which the geography of this or that area had been altered. Here are a few examples: Arthur Redford's account (1926) of migration in England during the first half of the nineteenth century set an example which was to be followed by many geographers. Different in character was T. W. Woodhead's description (1929) of the way in which the woodland of the southern Pennines in Norman times had been replaced by heather and bracken-covered slopes. E. W. Fenton (1937) was likewise writing about changes in vegetation in Scotland. C. S. Orwin's story of the reclamation of Exmoor was published in 1929, and his delightful essay on 'The taming of the wild' appeared in 1938 in the Orwins' book on open fields, another theme that appealed to geographers. Then in 1943 came an account of 'The reclamation of the waste in Devon, 1500–1800' by W. G. Hoskins which heralded the immense contribution that he was to make. His paper on the deserted villages of Leicestershire appeared in the following year, and this theme was later to be taken up by M. W. Beresford, again to our great advantage. The work of Hoskins and Beresford was soon to be appreciated as much by geographers as by historians. These various studies by historians and others differed in emphasis, but they all shared the common aim of demonstrating what has been called 'man's role in changing the face of the earth'.[50]

An end and a beginning

After the end of the Second World War many geographers returned to their posts, and prepared to take their ideas down from the shelf and to dust them after an absence of four or five years. When we looked back upon pre-war days we recalled a generation (our own) anxious to forge a method and to create an academic discipline. It is difficult today to realize how fragile our subject had been during the twenties and thirties, and how great was the need for intellectual underpinning. As late as 1934 it was said: 'Geography is far from having consolidated a definite position, British geography particularly so.'[51] But now, in 1945, there were reasons why we should take heart. In 1918 there had been geographical teaching of one kind or another in fourteen university departments. By 1945 this total had become twenty-nine, and, what is more important, most departments now had honours schools; the full-time academic staff numbered about one hundred and twenty.

I must not fall into the error of foreseeing with hindsight. We were certainly unaware of the changes that were almost upon us – of the great expansion in universities and, not least, in the teaching of geography within them.[52] Nor did we realize that the so-called 'quantitative revolution' was around the corner. But what we did know was that the work before the war had shown what immense riches awaited us – the Domesday Book, the Lay Subsidies, the Census Returns, the Tithe Returns, and other less known material such as Final Concords and the Probate Inventories, to name only a few. Moreover we were beginning to think more deeply about field systems and enclosure and about the growth of towns. We also had become aware of the possibilities of aerial photography.

Then, too, the work of economic historians had begun to make us realize that we were not the only inhabitants of the borderland between geography and history. Others also lived here, with their own methods of farming, and if we sometimes exchanged ideas, well there was no harm in that – maybe quite the reverse. It is probably true to say that our attitude was more relaxed than it had been in the 1930s. As we contemplated the variety of this borderland from the vantage point of a subject that had become more established, we began to feel it less necessary to insist on strict lines of demarcation. We were more inclined, at this fresh beginning, not to emphasize this or that dogma but to appreciate the commonsense statement of Sir John Clapham who had been so helpful at one stage in our development. He wrote in 1940: 'He is a very imperfect economic historian who is not also a tolerable geographer; and I cannot picture to myself a useful historical geographer who has not a fair working knowledge of economic history.'[53]

NOTES

[1] K. von Spruner, *Historisch-geographischer Hand-Atlas zur Geschichte der Staaten Europas* (Gotha, 1846); E. Wells, *An Historical Geography of the Old and New Testament* (London, 1840); C. Forster, *The Historical Geography of Arabia*, 2 vols (1844); W. M. Ramsay, *The Historical Geography of Asia Minor* (Royal Geographical Society, London, 1890). There were also

a number of French 'historical geographies' in the 1800s, e.g. V. Duruy, *Geographie historique du moyen âge* (Paris, 1842); L. Dussieux, *Géographie historique de la France ou histoire de la formation du territoire français* (1843).

2 H. F. Tozer, *Lectures on the Geography of Greece* (1873), p. v; the volume was dedicated to A. P. Stanley. See also A. P. Stanley, *Sinai and Palestine in Connection with their History* (1856).

3 C. L. Graves, *Life and Letters of Sir George Grove* (1903), 212–13; G. A. Smith, *The Historical Geography of the Holy Land* (1894).

4 J. Fairgrieve, *Geography and World Power* (various editions between 1915 and 1941). See also J. Fairgrieve and J. L. Myres, 'The content of historical geography' *Geographical Teacher*, 11 (1921–2), 40–3.

5 Ll. Rodwell Jones and P. W. Bryan, *North America: an historical, economic and regional geography* (1924); M. I. Newbigin, *The Mediterranean Lands: an introductory study in human and historical geography* (1924); V. Cornish, *The Great Capitals: an historical geography* (1923).

6 H. Butterfield, *The Present State of Historical Scholarship* (1965), 22.

7 H. B. George, *The Relations of Geography and History* (1901; 5th edn, 1924); E. A. Freeman, *The Historical Geography of Europe* (1880; 3rd edn, 1903); W. R. Shepherd, *Historical Atlas* (1911; 3rd edn, 1924); L. Febvre, *A Geographical Introduction to History* (1925); J. B. Bury, *A History of Greece* (1901; reprint of 1924).

8 H. C. Darby, 'The Fenland Frontier in Anglo-Saxon England', *Antiquity*, 14 (1934), 185–99; with E. Miller, Sections on 'Political history' in *V. C. H. Cambridgeshire and the Isle of Ely*, 2 (1948), 377–97; the publication of this was greatly delayed owing to the war.

9 H. C. Darby, 'The human geography of the Fenland before the drainage', *Geographical Journal*, 80 (1932), 420–35.

10 E. W. Gilbert, *British Pioneers in Geography* (1972), 143, 162; H. J. Mackinder, 'On the scope and methods of geography', *Proceedings Royal Geographical Society*, N.S., 9 (1887), 141–74.

11 F. J. Haverfield and G. Macdonald, *The Roman occupation of Britain* (Oxford, 1924), 89–124; J. N. L. Baker, 'The last hundred years of historical geography', *History*, N.S., 48 (1932), 206; and 'The development of historical geography in Britain during the last hundred years', *The Advancement of Science*, 8 (1952), 409.

12 H. J. Mackinder, 'Andrew John Herbertson', *Geographical Teacher*, 8 (1915), 144.

13 E. W. Gilbert, 'What is historical geography?' *Scottish Geographical Magazine*, 48 (1932), 132.

14 A. J. Herbertson, 'Regional environment, heredity and consciousness', *Geographical Teacher*, 8 (1915), 147–53.

15 J. F. Unstead, 'The meaning of geography', *Geographical Teacher*, 4 (1907), 28.

134   H. C. Darby

[16] J. F. Unstead, 'Geography and historical geography', Geographical Journal, 59 (1922), 55–9.
[17] Ll. Rodwell-Jones, 'Geography and the university', Scottish Geographical Magazine, 42 (1926), 77.
[18] P. M. Roxby, 'The scope and aims of human geography', Scottish Geographical Magazine, 41 (1930), 281.
[19] H. J. Mackinder, Britain and the British Seas (London, 2nd edn, 1915), 194–230; International Geographical Congress, Cambridge, 1928: Report of the Proceedings (1930), 310; and Geographical Journal, 78 (1931), 268.
[20] A. G. Ogilvie (ed.), Great Britain: essays in regional geography (1928), 423.
[21] International Geographical Congress, Cambridge: Report of Proceedings (1930).
[22] Premier congrès international de géographie historique, Compte-rendu des travaux du congrès, 45–7, 114–6; and Mémoires, 212–26 (1931).
[23] 'What is historical geography?' Geography, 17 (1932), 39–45.
[24] E. W. Gilbert, 'What is historical geography?' Scottish Geographical Magazine, 48 (1932), 129–36.
[25] W. G. East, 'A note on historical geography', Geography, 18 (1933), 282–92.
[26] J. H. Clapham, 'Economic history as a discipline', Encyclopaedia of the Social Sciences, 5 (1931), 327–30.
[27] Marc Bloch, 'En Angleterre: l'histoire et le terrain', Annales d'histoire économique et sociale, 9 (1937), 208–10.
[28] H. C. Darby, 'An historical geography of England: Twenty years after', Geographical Journal, 126 (1960), 147–59.
[29] E.g. H. C. Darby, 'The problem of geographical description', Transactions and Papers of the Institute of British Geographers, 30 (1962), 1–14; H. C. Prince, 'Real, imagined and abstract worlds of the past', Progress in Geography, 30 (1971), 4–86; D. W. Moodie and J. C. Lehr, 'Fact and theory in historical geography', The Professional Geographer, 28 (1976), 132–5.
[30] W. G. East, 'Land utilization in England at the end of the eighteenth century', Geographical Journal, 89 (1937), 156–72; and 'Land utilization in Lanarkshire at the end of the eighteenth century', Scottish Geographical Magazine, 53 (1937), 89–110.
[31] H. C. Darby, Domesday England (Cambridge, 1977), 376; F. Walker, Historical Geography of Southwest Lancashire before the Industrial Revolution (1939); H. C. Darby, The Medieval Fenland (1940).
[32] E. G. Bowen, Wales: a study in geography and history (1941).
[33] C. B. Fawcett in 'What is historical geography?' Geography, 17 (1932) 40.
[34] J. F. Unstead, 'Geography and historical geography', Geographical Journal, 59 (1922), 55–9.
[35] H. J. Mackinder in a discussion of a paper by S. W. Wooldridge and D. J. Smetham at the R.G.S., Geographical Journal, 78 (1931), 268.

[36] E. C. Willatts, 'Changes in land utilization in the south-west of the London Basin, 1840–1932', *Geographical Journal*, 82 (1933), 515–28; H. C. K. Henderson, 'Our changing agriculture: The Adur Basin, Sussex, 1780–1931', *Agriculture*, 43 (1936), 625–33; Arthur Geddes, 'The changing landscape of the Lothians, 1600–1800', *Scottish Geographical Magazine*, 54 (1938), 129–43; J. H. G. Lebon, 'The face of the countryside in central Ayrshire during the eighteenth and nineteenth centuries', *Scottish Geographical Magazine* 62 (1946), 7–15; H. C. Darby, *The Draining of the Fens* (1940).

[37] S. J. Jones, 'The historical geography of Bristol', *Geography*, 16 (1931), 175–86; J. B. Mitchell, 'The growth of Cambridge' being chapter 12 (162–80) of H. C. Darby (ed.), *The Cambridge Region* (1938); S. G. E. Lythe, 'The origin and development of Dundee: a study in historical geography', *Scottish Geographical Magazine*, 54 (1938), 344–57; G. de Boer, 'The evolution of Kingston-upon-Hull', *Geography*, 31 (1946), 139–46; R. M. Rudmose-Brown, 'Sheffield; its rise and growth', *Geography*, 21 (1936), 175–84; W. G. East, 'The historical geography of the town, port and roads of Whitby', *Geographical Journal*, 80 (1932), 484–97; E. W. Gilbert, 'The growth of inland and seaside health resorts in England', *Scottish Geographical Magazine*, 55 (1939), 16–35; H. Ormsby, *London on the Thames*, 1923; 2nd edn (1928); Ll. R. Jones, *The Geography of London River* (1931).

[38] F. T. Baber, 'The historical geography of the iron industry of the Forest of Dean', *Geography*, 27 (1942), 54–62. T. W. Birch, 'Development and decline of Coalbrookdale Coalfield', *Geography* 19 (1934), 114–26. A. E. Smailes, 'The development of the Northumberland and Durham Coalfield', *Scottish Geographical Magazine*, 51 (1935), 201–14. R. P. Beckinsale, 'Factors in the development of the Cotswold woollen industry', *Geographical Journal*, 90 (1937), 349–62.

[39] R. A. Pelham, 'The immigrant population of Birmingham, 1685–1726' *Transactions of the Birmingham Archaeological Society*, 61 (1937), 45–80; A. E. Smailes 'Population changes in the colliery districts of Northumberland and Durham', *Geographical Journal*, 91 (1938), 220–32; H. C. Darby, 'The movement of population to and from Cambridgeshire between 1851 and 1861', *Geographical Journal*, 101 (1943), 118–25; T. H. Bainbridge, 'Cumberland population movements, 1871–81', *Geographical Journal*, 108 (1946), 80–4.

[40] J. H. Clapham, *An Economic History of Modern Britain*, vol. 1, chapter 1 for 1820; vol. 2, chapter 2 for 1886–7 (1926; 1932); G. M. Trevelyan, *England under Queen Anne* (1930), the first four chapters of which were reprinted as *The England of Queen Anne* (1932); G. D. H. Cole, Introduction to *Daniel Defoe's tour thro' the whole island of Great Britain* (1927); G. D. H. Cole, 'Defoe's England', *Persons and Periods*, 31–38 (1938); D. Ogg, *England in the reign of Charles II* (Oxford, 1934), vol. 1, chapter 2; A. L. Rowse, *The England of Elizabeth* (1951), 66; S. T. Bindoff, *Tudor England* (1950), 9.

41 W. H. B. Court, *The Rise of the Midland Industries, 1600–1838* (1938), p. v; J. D. Chambers, *Nottinghamshire in the Eighteenth Century* (1932), especially chapters 4, 6 and 7; T. S. Willan, *River Navigation in England, 1600–1750* (1936); T. S. Willan, *The English Coasting Trade, 1600–1750* (1938); W. Rees, *Historical Map of South Wales and the Border in the XIVth century* (published by the University of Wales, 1933, but printed by the Ordnance Survey); W. Rees, *South Wales and the March, 1284–1415* (1924).

42 A. Zimmern, *The Greek Commonwealth: Politics and economics in fifth-century Athens* (1911; 4th edn, 1924); J. L. Myres, *Geographical History in Greek Lands* (1953), being a collection of some of his works.

43 M. P. Charlesworth, *Trade Routes and Commerce of the Roman Empire* (1924).

44 C. H. Pearson, *Historical Maps of England during the First Thirteen Centuries* (1869); J. R. Green, *The Making of England* (1881); E. Guest, *Origines Celticae, and other Contributions to the History of Britain*, 2 vols (1883); J. P. Williams-Freeman, *An Introduction to Field Archaeology as Illustrated by Hampshire* (1915).

45 O. G. S. Crawford, *Said and Done: the autobiography of an archaeologist* (1955), 216.

46 O. G. S. Crawford, *Archaeology in the Field* (1953), 41–2.

47 C. W. Phillips, 'The special archaeological and historical maps published by the Ordnance Survey', *Cartographic Journal*, 2 (1965), 27–31; J. B. Harley, *Ordnance Survey Maps: a descriptive manual* (1975), 151–7.

48 Cyril Fox, *The Archaeology of the Cambridge Region* (1923); Cyril Fox, *The Personality of Britain* (1932); Carl Sauer, 'The personality of Mexico', *Geographical Review*, 31 (1941), 353–64. See also Cyril Fox, 'Reflections on the Archaeology of the Cambridge region"', *Cambridge Historical Journal*, 9 (1947), 1–21.

49 R. L. Sherlock, *Man as a Geological Agent* (1922), 14; and 'The influence of man as an agent in geographical change', *Geographical Journal* 61 (1923), 258–73.

50 A. Redford, *Labour Migration in England, 1800–50* (1925); T. W. Woodhead, 'History of the vegetation of the southern Pennines', *Journal of Ecology*, 17 (1929), 1–34; E. W. Fenton, 'The influence of sheep on the vegetation of hill grazings in Scotland', *Journal of Ecology*, 25 (1937), 424–30; C. S. Orwin, *The Reclamation of Exmoor Forest* (1920); C. S. and C. S. Orwin, *The Open Fields* (1938), 15–20; W. G. Hoskins, 'The reclamation of the waste in Devon, 1550–1800', *Economic History Review*, 13 (1943), 80–92; W. G. Hoskins, 'The deserted villages of Leicestershire', *Transactions of the Leicestershire Archaeological Society*, 22 (1944–5), 242–64; M. W. Beresford, 'The deserted villages of Warwickshire', *Transactions of the Birmingham Archaeological Society*, 66 (1950), 49–106; W. L. Thomas (ed.), *Man's Role in Changing the Face of the Earth* (1956).

51 In an obituary of 'Marion Isabel Newbigin', *Scottish Geographical Magazine*, 50 (1934), 331–3.

52 H. C. Darby, 'Academic geography in Britain: 1918–46', *Transactions of the Institute of British Geographers*, 8 (1983), 14–26.

53 J. H. Clapham in the Editor's Preface to H. C. Darby, *The Draining of the Fens* (1940), ix.

10 Physical geography in the universities, 1918–1945

J. A. STEERS*

What was meant by physical geography? In the years following 1918 most people probably included in their answer what was in the then standard book on the subject, *Physical Geography*, by Philip Lake, which was first published in 1915. It was in three sections – elementary meteorology and climatology, oceanography and landforms. It was finally printed in 1958, having been considerably enlarged and in part re-written. Nevertheless, for about half a century it was used both for first-year work at universities and for sixth-form work in schools. But during that time the subject had expanded greatly, and in the 1930s several more specialist books were written and were in general use. But advanced courses in physiography or geomorphology could not be attempted without the reading of many papers, largely, but not wholly, on fieldwork and research into the origin of landforms and related matters. These appeared in scientific journals, mainly of geology and geography. Although some courses still required a knowledge of climate and meteorology and possibly of oceanography, these subjects at an *advanced* level were basically the field of physicists, chemists and biologists. I have stressed advanced level; a general knowledge of both oceanography and climatology with meteorology was doubtless taught in several departments, and at Cambridge, until recent changes, reasonably detailed lectures were given in both subjects. The original intention to make oceanography and climatology a special subject in Part II of the Tripos

* James Alfred Steers, C.B.E. (b. 8 August 1899) graduated in geography in the University of Cambridge in 1921. He was appointed as a departmental demonstrator in Cambridge in 1922 and a University demonstrator in 1926. In 1925 he was elected a Fellow of St Catharine's College, Cambridge, where he was subsequently Dean, Tutor and President. From 1927 until 1949 he was a University Lecturer in Geography; in 1949 he became Professor of Geography, on his retirement he became Emeritus Professor. He was an Emeritus Fellow of St Catharine's College, Cambridge. He died on 10 March 1987.

was short-lived since very considerable ability in mathematics, physics, chemistry and biology is essential in any deep study of these subjects. (I think I am right in saying that G. Manley was the only candidate who took these papers.) How advanced Manley's lectures were outside Cambridge I do not know, but in Cambridge he did not go above the heads of average geography students. I cannot comment on A. A. Miller's and P. R. Crowe's teaching at Manchester and Glasgow. Geomorphology, however, became increasingly a study for geographers and W. M. Davis and his disciples in the USA and the UK and several other countries set the example that was generally followed. This produced many good papers and was a popular field of study. It was, however, often too theoretical and it omitted to emphasize what may broadly be referred to as the ecological approach which is extremely important in many landform studies. T. H. Huxley gave a more realistic picture since it included, for example, the effect of the plant cover and also encouraged the student to give more consideration to processes and, in fact, to all natural factors that play a part, and often an important part, in the evolution of the landscape. This was also demonstrated by E. de Martonne in his three volumes, *Traité de géographie physique*. Moreover, measurement of changes and of the speed with which processes act became far more significant.

When I was invited to write this chapter I sought information from all geography departments of universities that were active between 1918 and 1945. Since 1945 several new universities have been founded and geography has become a subject studied by a large number of students. It has also changed its nature considerably. There is no need to consider these changes here; in fact they have only become fully significant since about 1960 or even a little later.

Before 1918 there were, in this country, no trained geographers. Geography had to be established as a teaching subject by people who had been trained in some other branch of learning, but who nevertheless realized the value of geography as a university discipline in training students for posts in business, in certain professions, in teaching, and in the period with which we are concerned geographers began to hold important positions in the Colonial Service as administrators or surveyors. It was also inevitable that at first a large proportion of students became school teachers, but once this demand was satisfied, more and more sought, and found, openings in business, in various branches of the higher Civil Service and elsewhere.

The introduction of the new subject into universities often gave rise to a good deal of discussion. Between about 1910 and 1930 it occasionally provoked not only discussion, but also some faculty opposition which was little more than inherent conservatism and traditional opposition to innovative change. Sometimes, however, it was frustrating.

It is now nearly seventy years since the end of the First World War, and forty-eight since the Second World War began. Relatively few geographers remember the state of geography in universities between the wars and fewer still recall the names of many who were then active. It is for this reason that this section is included. It also indicates how few geographers taught physical geography.

England

(a)   *Oxford:* H. O. Beckit was head of the School of Geography from 1919 to 1931. Although interested in geomorphology he did not write much, but was responsible for good work in the Oxford region. W. G. Kendrew, well known for his books on climate, was a classicist. Climatology was his hobby and he organized a small meteorological station in his garden on Cumnor Hill. The chair was founded in 1932 and Colonel Kenneth Mason, primarily a surveyor, was its first occupant.

(b)   *Cambridge:* Geography was established in 1914 and from the first had a strong physical basis. P. Lake was appointed Reader when the Tripos was instituted in 1919. In 1927 he was succeeded by F. Debenham, a geologist and surveyor who, in 1931, became the first professor of geography. I returned to Cambridge in 1922 as a departmental demonstrator and became a University demonstrator in 1926 when the new statutes came into force, both at Oxford and Cambridge. I was promoted to a University lecturership in 1927. W. V. Lewis was made a University demonstrator in 1933, but was not promoted to a lectureship until after the war (1946).* G. Manley held a demonstratorship in 1939, and left as a lecturer in 1948. During his residence in Cambridge much of his time was given to the University Air Squadron. Debenham also became first Director of the Scott Polar Research Institute and encouraged interest in polar expeditions.

(c)   *Durham (including Newcastle):* The teaching of geography began in 1928 at Durham, and G. Manley was in charge until 1938. He was succeeded by L. Slater, who left in 1939 for war service.

* University demonstratorships were held for a maximum of seven years, but when war broke out anyone likely to become a lecturer was reappointed annually as a demonstrator until after the war ended. In 1939 Lewis had not reached his maximum tenure as a demonstrator.

F. Peel was at Newcastle for part of the session 1939–40. Durham was at this time a federal university and included King's College at Newcastle. Geography was taught at both places on a more or less common syllabus. The Newcastle department was established in 1928. Peel returned to Newcastle in 1945–6.

(d)   *London:*

(i) University College London: No physical geography was taught in the department between 1913 and 1945, but (see below, p. 144) great help was given by geologists.

(ii) King's College and London School of Economics: Until 1947 geography was a sub-department of geology. In that year S. W. Wooldridge was appointed professor and the department of geography became independent. The decision to establish a joint school was taken in 1921 though it was not fully operative until 1930. Professor Rodwell Jones was in charge at L.S.E. and Wooldridge at King's. From 1926, when L. Dudley Stamp was appointed to a readership at L.S.E., the connection grew much closer.

(iii) Birkbeck: From 1918 to 1945 physical geography was taught by geographers. Until 1920 the course was that for the external London degree but in 1920 the College became an internal school of the university and J. F. Unstead was appointed to a foundation chair in 1922. In 1930 he was succeeded by Eva G. R. Taylor and from 1944 to 1947 S. W. Wooldridge held the professorship.

(iv) Queen Mary College: From 1914 to 1947 all physical geography was taught by H. G. Smith who was head of the joint department of geology and geography.

(v) Bedford College: Before 1920 geography and geology formed one department under Dr Catharine Raisin, a geologist. After 1921 B. Hosgood took charge of geography and L. Hawkes of geology. G. Manley succeeded Miss Hosgood in 1945.

(e)   *Manchester:* Professor H. B. Rodgers told me that when he was a student (1941–2) he was taught the barest minimum of physical geography. Up to that time Professor H. J. Fleure had determined the syllabus: his interest was in the facts of the environment and less with the processes that gave rise to it. Until after the war geomorphology had no place in the syllabus. Some climatology was taught for short periods by N. Pye, F. K. Hare and R. Miller before the war; Pye also taught some after the war.

(f)   *Liverpool:* The teaching of geography began in 1886, and there was a department in 1909. The syllabus included some physical geography, but the first appointment of a member of staff to teach it was in 1946 when R. K. Gresswell was appointed a part-time lecturer.

(g)   *Leeds:* C. B. Fawcett was the first lecturer appointed in geography (1919) and an honours course was established in 1920. Two papers in the final examination were on geomorphology and climatology. A department was inaugurated in 1928 and A. V. Williamson was elected professor in 1945. In 1946 Miss Anne Priestly was appointed to teach geomorphology.

(h)   *Sheffield:* Professor R. N. Rudmose Brown and Dr Alice Garnett both lectured on physical geography which was a significant element in the syllabus in the inter-war years.

(i)   *Bristol:* Formal geographical teaching began in 1920; it became an independent subject in 1925. O. D. Kendall taught most of the physical geography as well as surveying. W. W. Jervis, who became first professor in 1933, gave some lectures in climatology and F. G. Morris was interested in erosion processes and hydrology.

(j)   *Reading:* H. N. Dickson was professor of geography from 1907 to 1920. The department was reorganized in 1925–6, in which year A. A. Miller joined the staff. He became professor in 1943.

(k)   *Nottingham:* Until 1934, when the joint department of geology and geography was split, geography depended wholly on Professor H. H. Swinnerton who continued to teach until 1945. He was succeeded as professor by K. C. Edwards. It was not until 1951 that geomorphology was taught systematically by Cuchlaine A. M. King, who later became a professor.

(l)   *Southampton:* C. B. Fawcett was the first lecturer in geography. He was succeeded by W. H. Barker in 1921. O. H. T. Rishbeth followed as reader and was appointed professor in 1926. The chair was vacant between 1938 and 1954 when F. J. Monkhouse, a physical geographer, was appointed.

(m)   *Hull:* A department was established in 1928 and H. King was appointed lecturer in charge. He was solely responsible for teaching until G. H. T. Kimble arrived in 1931; he stayed until 1936. There were several changes during the war.

(n)   *Exeter:* Geography courses for teachers were offered between 1918 and 1945; these included some physical geography, and emphasis was placed on the physical basis of the subject as a whole. Until 1927 geography was twinned with geology. A. W. Clayden was responsible for the teaching of geomorphology; he retired in 1920 but was 'visiting director' to the department until his death in 1944.

In 1927 W. S. Lewis became the first occupant of the Reardon Smith chair.

(o)   *Leicester:* P. W. Bryan was appointed lecturer in 1922 and covered all that was required for the London external degree until 1947 when his first assistants, J. N. Jennings and R. Millward, were appointed.

It should be noted that the University colleges of Reading, Nottingham, Southampton, Hull, Exeter and Leicester, before they became independent institutions, prepared students for the External London degree.

Wales

(a)   *Aberystwyth:* Geography was dominated by Professor H. J. Fleure whose chair covered both geography and anthropology. There seems to have been no formal teaching in physical geography, but Fleure's lectures in regional, economic and historical geography gave his students some knowledge of that aspect of geography in a regional setting. C. Daryll Forde, who succeeded him in 1930, appears to have followed similar lines.

(b)   *Swansea:* Geography was taught from 1920, the date of the inauguration of the college (see below, p. 145). In the early days S. W. Rider, a master at Gowerton secondary school, gave very successful lectures on Saturday mornings to geographers.

(c)   *Cardiff:* Before 1939, F. J. North, a geologist, was alone responsible for any teaching in physical geography.

Scotland

(a)   *St Andrews:* A general undergraduate M.A. course in geography, which contained some physical geography, began in 1935–6. The subject developed slowly, and it was not until after 1945 that there was a department.

(b)   *Glasgow:* From 1919 to 1945 A. Stevens was lecturer in charge of the department. He became professor in 1947. He had been with Shackleton in the Antarctic. P. R. Crowe introduced some climatology and elementary meteorology.

(c)   *Aberdeen:* J. MacFarlane was in sole charge of a small department from 1918 to 1945. He was mainly interested in economic geography but he also introduced some physiography and climatology.

A. C. O'Dell was the first professor; his successor, K. C. Walton, developed the teaching of geomorphology.

(d)   *Edinburgh:* A little physical geography was taught between the wars when G. G. Chisholm was appointed in 1915–19. A. G. Ogilvie (see below) succeeded him and became the first professor in 1931. He developed the physical side, which was greatly stimulated when D. L. Linton joined the staff in 1929, and especially when he returned after the end of the Second World War.

Northern Ireland

Belfast; The Queen's University: From 1915 to 1945 both teaching and research in geography took place in the department of geology. J. K. Charlesworth, the first professor of geology made geography a sub-department, but was never given the title of professor of both subjects. Estyn Evans became head of the geography school in 1928, and an honours school was founded in 1931, but was still a joint school with geology. Evans became reader in 1944 and professor in 1945.

The help given by geologists

Between the wars relatively little physical geography, especially geomorphology, was taught by members of the staffs of geographical departments. It is easy to forget how much help was sometimes required to establish geography as a university discipline. Geologists were often the people who understood the need best, and this is a good opportunity to record what they did. The nature of their help varied considerably, of course, from place to place.

The several colleges in London were much helped, especially in geomorphology. E. J. Garwood at University College was a strong supporter. Until 1945 no physical geography was taught in the department, but Garwood and his successor, S. E. Hollingworth, were not only immensely helpful, but because of their own distinction in the college did much to establish the present standing of the subject in the department. They both taught well and took a real interest in the development of the department. They also took students into the field. At King's College, W. T. Gordon, the first professor of geology, saw the need for the re-birth of geography. The subject had been taught at L.S.E. since 1895 by H. J. Mackinder, and the joint school was achieved in 1930. Gordon's pioneer work led to one of his pupils, S. W. Wooldridge, becoming professor at King's after he had held the chair at Birkbeck for three years. L. Dudley Stamp was also a King's man. Although he is best known for his work in other branches of the subject, he published several important papers on geomorphology and also a well-known text book on geology, a book which owed much to Gordon's lectures. At Bedford College Professor L. Hawkes taught physical geography before and after the department was founded.

Two particularly good examples of the help given by geologists to geography occurred at Swansea and Bristol. At Swansea, A. E. Trueman was in charge of geology, and he had very wide interests. His book, *The Scenery of England and Wales* (1939), later published as a Penguin book and now revised by J. B. Whittow and J. R. Hardy, had a great influence and showed his own interest in physiography. In 1947 Trueman and I were both members of the Wild Life Conservation Special Committee which led to the formation of the Nature Conservancy. It was then that I realized the genuine interest Trueman had in geography and in the establishment of sites of physiographical interest. He was succeeded by T. N. George who had similar interests and helped greatly in establishing the flourishing department of geography at Swansea. George left in 1941 for Glasgow and his influence in geography continued at that University. Trueman left Swansea for Bristol where he followed S. H. Reynolds. Both helped to put geography on a firm footing. Trueman's later influence was in a different sphere – as Chairman of the University Grants Committee.

Among geologists who played important parts in helping new departments of geography were K. S. Sandford at Oxford, G. Hickling at Newcastle, C. Lapworth at Birmingham, (Sir) William Pugh (who later became Director-General of the Geological Survey) at Manchester, P. G. H. Boswell at Liverpool, H. L. Hawkins at Reading, H. H. Swinnerton at Nottingham, A. W. Clayden at Exeter, T. J. Jehu at Edinburgh and J. K. Charlesworth at Belfast. It is nowadays all too easy to forget how many departments reached full independent status only after the war. It is in quite recent years that staff numbers have grown so that specialist teaching in most universities can be given in any branch of the subject.

Did geologists give an undue emphasis to structure? This question has been raised from time to time. I can only express a personal view – that, in general, a very fair balance between structure and other factors in the evolution of landscape was maintained by most geologists who lectured to geographers on geomorphology. Some were keenly interested in the evolution of scenery. If any criticism is valid, I should say that the effects vegetation had in certain areas (especially coastal, lacustrine, dune and deltaic) were the most likely to be omitted.

The beginning of measurements and the analysis of physical processes in physical geography

Cambridge was, I think, the first department to lay great stress on physical geography. Three heads of department had held office, two for very short periods, before P. Lake was made Reader in 1919. He was succeeded by Debenham and Steers, both of whom were largely concerned with the physical side of the subject. Debenham was also much interested in survey. It was largely his influence that encouraged Steers and W. V. Lewis to apply measurements and to make what may be called numerical experiments in their fieldwork. This began in a simple way. Lewis, who was closely associated with Brathay almost from its foundation, suggested and, with the help of Brathay students, carried through, the sounding of a number of lakeland tarns, a task which helped greatly in explaining their origins. On the coast several simple, but definitely helpful, experiments were made on the rate of direction of movement of marked pebbles. After the war this work was carried much further by means of experiments with radioactive material and also by diving and noting the movements – or sometimes the total lack of movement – of pebbles that had been placed by divers in selected and carefully surveyed sites.

These experiments were made on Scolt Head Island, where A. T. Grove also carried out some very useful work on the measurements and the explanation of changes of beach profiles. Cliff erosion was also measured and considered along all the coasts of Norfolk and Suffolk and also in relation to the over-rolling of, for example, Blakeney Spit and Orford Ness. The accretion of mud and blown sand on salt marshes was measured for a period of twelve years beginning in 1935. The reasons for the different rates of accretion on several marshes were analysed. In the years following the end of the war C. Kidson and A. P. Carr (1959), both officers in the Nature Conservancy, made some valuable measurements of pebble drift on and off the southern end of Orford Ness, and when the Hydraulics Research Station at Wallingford opened work of this type was greatly improved and extended. In Lincolnshire Cuchlaine A. M. King and F. A. Barnes (1964) carried out valuable experiments on drift and on spit and bar formation on the coast of Lincolnshire.

It is relevant here to add that, long before the so-called 'quantitative revolution', considerable time in several departments was devoted to the teaching of surveying and map projections. These studies now seem to be out of fashion but they demanded, or could demand, according to the standards required by different teachers, considerable mathematical ability. What is more they were of real value to the students, especially in certain kinds of fieldwork and in understanding the limitations of atlas, topographical and other maps. This chapter is not concerned with modern developments. I would only venture to say that maps are essential tools of the geographer and that he should be competent to make his own in many kinds of fieldwork; have access, preferably in the department in which he is studying, to maps of all kinds and types; and be able to judge the limitations of particular projections and types of maps.

In all experiments – including some of those noted above at Scolt – we owe much to N. C. Flemming. Despite the great handicap he suffered as a result of a serious motor accident in Turkey, he, fortunately, was still able to help in laboratory and theoretical work.

Research in geomorphology and physiography

(a) Historical geomorphology: S. W. Wooldridge soon became a leader, and a distinguished one, in this aspect of the subject. One of his earliest papers, on the Mole gap, was written in collaboration with A. J. Bull, an older man and a geologist (Wooldridge and Bull 1925). Wooldridge in the 1920s and 1930s made important studies in the Thames basin. He was largely concerned with its history in Tertiary and later times, and also on the origin of the 200-foot platform (Wooldridge 1926; 1927; 1932). In 1936 he and J. F. Kirkaldy discussed river profiles and chronology in south-east England (Wooldridge and Kirkaldy 1936), and in 1938, with D. L. Linton, he wrote a paper on the influence of the Pliocene transgression on the geomorphology of the same region (Wooldridge and Linton 1939). This was the first of several joint papers with Linton; the others appeared after the war. It was in 1939 that their well-known *Structure, Surface and Drainage in South-east England* was published. Linton, while a lecturer at Edinburgh wrote several papers on Scottish rivers, following his earlier work on those in Wessex (Linton 1932; 1933a; 1934; 1940). In 1938 A. A. Miller wrote on river development in southern Ireland. Previously he had published significant papers on the meanders of the Herefordshire Wye, the pre-glacial erosion surfaces around the Irish Sea, and the 600-foot platform in Pembrokeshire and Carmarthenshire (Miller 1935; 1937; 1938; 1939). C. F. W. R. Gullick (1936) and W. G. V. Balchin (1937) made similar studies on the platforms along parts of the Cornish coast and J. Hanson-Lowe (1938) analysed those of the Channel Islands. R. O. Jones (1939) studied the evolution of the

Tawe and Neath drainage. J. F. N. Green *et al.* (1934) added much detail to the history of the Mole, and in 1941 Green discussed the high platforms of east Devon (Green 1934, 1941). Lake (1934) wrote an interesting account of the possible connection of the Welsh rivers with the Thames, and W. V. Lewis (1945) dealt with the significance of nick points in relation to the curve of water erosion.

(b) Glaciated Regions: There was an increasing interest taken by geographers in glaciated landscapes, but nearly all of it bore fruit after the Second World War. W. V. Lewis (1938a) wrote a paper on a melt-water hypothesis of cirque formation which gave rise to a good deal of discussion. Linton (1933b) wrote on the Tinto glacier and glacial features in Clydesdale. The joint work of A. R. Dwerryhouse and Miller (1930) on the glaciation of Clun and Radnor forests is noteworthy, while S. E. Hollingworth, T. N. George and A. J. Bull all published work of marked geomorphological nature. (For work done by those who went on Polar Expeditions between the wars, see pp. 151–2).

(c) Arid Regions: In these regions academic geographers did little work before the war with the single exception of R. F. Peel. He accompanied R. A. Bagnold in 1938 to south-west Egypt. Bagnold, who was a Major in the Royal Engineers, travelled widely in the Sahara and became so fascinated with the problems it presented that he resigned his commission and carried out some distinguished research work in a laboratory in Imperial College. He published his well-known book, *Physics of Blown Sand and Desert Dunes*, in 1941 and in the Second World War rejoined the Army and became a Brigadier, and made great practical use of his knowledge of the Sahara and his laboratory work.

His writings, and Peel's interest, inspired several geographers who have written on desert landscapes since the war. It is also noteworthy that K. S. Sandford of Oxford, who gave much of his time to teaching geographers, was also an authority on arid regions. Miss Caton Thompson, a former Fellow of Newnham College and later an Hon. Litt.D., attended some geomorphological lectures in the department of geography in the 1920s, and her work on the northern Faiyum in 1924–6 included much of physiographical interest.

(d) Climatology: Gordon Manley held a unique position as a geographer. His interest and important work on meteorology and climatology began in the late 1920s and was probably helped by his visit to the Arctic with (Sir) James Wordie in 1925. His first published paper on climate (on the weather of the High Pennines) appeared in 1932. It was the first of one hundred and forty papers and notes, twenty-two of which had been written and published by the end of 1939. He wrote one well-known book, *Climate and the British Scene*, which received its fifth printing in 1972. Apart from one year in the Meteorological Office, he was in university work until his retirement in 1968. He continued to be very active in research and his last published paper appeared in 1978, two years before his death.

(e) Coastal Studies: The first edition of D. W. Johnson's *Shore Processes and Shoreline Development*, published in the USA in 1919, had a considerable influence on the future of coastal research. It was the first book to show both the scope and wide interest of the subject. In Britain a number of papers, and two or three books, had been written on the coast. Several were the work of geologists and engineers concerned with sea defences. The works of Sir John Coode, Sir Joseph Prestwich, F. P. Gulliver, J. B. Redman, W. H. Wheeler, Vaughan Cornish and Miss E. M. Ward are especially important. There were also many pages, some of considerable value, in those Memoirs of the Geological Survey which covered coastal areas. In 1914 and 1923 A. G. Ogilvie published two interesting papers on shingle and sand formations in the Moray Firth.

Lake in his advanced lectures touched on some of these papers and problems and introduced the subject at a high level. In the academic year 1922–3 the present writer returned to Cambridge as a departmental demonstrator, and Lake suggested that he should attempt fieldwork on Orford Ness. A good deal was done in the Easter and Long Vacation of 1923 which led to papers published in 1925 and 1926 (Steers 1926). The East Anglian coast soon proved to be of great interest throughout its entire length. This is particularly true of the coast of north Norfolk. Professor F. W. Oliver and Professor (Sir) Edward Salisbury of University College London had studied the botany and ecology of Blakeney Point, but little had been written about the nature and origins of the Point. It was a property of the National Trust, and Scolt Head Island soon became another Trust property. Scolt had attracted little attention except from a few botanists and ornithologists. It soon, however, became an active centre for the Cambridge department of geography, and several students who later became academics, undertook their first fieldwork there. O. D. Kendall was the first to make a physiographical map of the island, and later, when he was at Bristol, he published material on the coast of Somerset (Kendall 1936). A few years later R. F. Peel surveyed the island in much greater detail; and a third map, based on a survey by P. Haggett and R. J. Small, was greatly amplified by the insertion of numerous creeks and other features from vertical air photographs. Later surveys have been made by the officers of the Nature Conservancy Council. The island is one of the finest coastal nature reserves in any country. In 1924 a book, *Scolt Head Island*, was published to which fourteen people contributed. V. J. Chapman, for many years professor of botany at the University of Auckland, wrote his Ph.D. thesis (Cambridge) on the ecology of the island; the substance of this appears as Chapter VIII in the book. A second edition, enlarged and re-written only in part by the same authors that contributed to the 1934 edition, was published in 1960 (Steers 1960).

W. V. Lewis visited Scolt on several occasions and his interest in coasts largely sprang from these visits. He wrote four significant papers on coastal features: 'The effects of wave incidence on the configuration of shingle beaches' (1931), 'The formation of Dungeness' (1932), 'The evolution of shoreline curves' (1938b) and 'Past sea-levels at Dungeness' (1940, with W. G. V. Balchin). W. W. Williams did some important work on beach profiles and beach bars. This was done partly in a wave tank and partly on the coast, and was largely related to the beach landings in Normandy in 1944 (Williams 1947). Professor Cuchlaine A. M. King's detailed work on the Lincolnshire coast was done after the war, but was preceded by two papers by H. H. Swinnerton (1931; 1936): one on the post-glacial deposits of that coast, and the other on the physical history of east Lincolnshire, in which he dealt with the coast in some detail.

The present writer worked on the Culbin Sands in 1937, and on the sand and shingle formation of Cardigan Bay in 1938–9. He was also responsible for various papers on the East Anglian coast, among which were two or three on the rate of sedimentation on the salt marshes of Scolt Head Island between 1935 and 1947. In 1928 and 1936 he made extensive voyages within the Great Barrier Reefs of Queensland in order to study coastal formations on the mainland and high islands, and particularly the sand cays and low wooded islands. In 1928 Michael Spender of Oxford accompanied him as a surveyor and E. C. Marchant, a Cambridge geographer of the early 1920s, joined at Cooktown and helped greatly in the survey work. For some weeks Marchant remained with Spender at Low Isles (the headquarters of the main expedition) to assist in making the large-scale and very detailed ecological and physiographical maps of these islands. In 1936 F. E. Kemp, a Cambridge geographer, accompanied me on a second expedition in which we mapped many low wooded islands and covered the coast from Brisbane to Cape Direction.

In 1939 I joined Professor V. J. Chapman (see above) and others in an investigation of the sand cays and low islands around Jamaica. In this visit the Morant Cays and the Bogue Islands were included. Although war broke out while we were in the Caribbean, there was time also to examine the structure of the Palisadoes, the spit which encloses Kingston Harbour.

(f) Three other items deserve mention. In 1941 O. T. Jones, professor of geology at Cambridge, was helped by W. V. Lewis in making his observations on the water level in the Breckland meres (Jones and Lewis 1941). Shortly before the war J. N. Jennings began his work on the origin of the Broads. Later he collaborated with Dr J. M. Lambert, a botanist at Southampton. The work was resumed after the war and C. T. Smith of Cambridge (then at Leicester) contributed a valuable account of the historical approach to the problem. Their joint work was published as a volume of the Royal Geographical Society's research series (Lambert *et al.* 1960).

Mention must also be made of the work on limestone topography which began during the war with Marjorie M. Sweeting's thesis on the limestone areas of Yorkshire. Since the war she has been responsible not only for a great deal of work of her own, but also in inspiring others to follow her lead in many parts of the world.

In 1940 Miss M. A. Arber, a geologist, published her paper on 'The coastal landslips of south-east Devon' (Arber 1940). Since then she has extended her work and others, especially Professor J. N. Hutchinson (1936) an engineer interested in coastal evolution, and three geographers, D. R. Brunsden, D. K. C. Jones (1976) and E. Derbyshire *et al.* (1979) have added considerably to our knowledge in more recent years.

(g) Long Vacation expeditions and certain Polar Expeditions: Occasional reference has been made to expeditions in which geographers have played a part. Attention must now be called to the growth of Long Vacation expeditions and also to a few major ones. Many originated at Oxford and Cambridge and somewhat later extended to some other universities. Since 1945 the number of expeditions has greatly increased and almost all universities play a part in them. Nearly all were, and are, made up of small groups chosen from different faculties so that their work, and reports, can be more comprehensive and of greater interest to many readers. Some accounts appear at length in the *Geographical Journal* which also lists all expeditions. Geographers often play an important part in them and, as will be seen, some have later distinguished themselves in academic life.

L. Slater went as surveyor to the Oxford Expedition to British Guiana led by R. W. G. Hingston (Hingston 1930). On the British Arctic Air Route Expedition H. G. Watkins and August Courtauld, both of whom read for the now defunct Pass Degree at Cambridge, taking geography as one of their subjects, were accompanied by A. Stephenson, who also worked at Kangerlugsuak and Mount Forel. Brian Roberts, who took his degree in geography at Cambridge and later worked partly at the Scott Polar Research Institute and the Colonial Office, and W. V. Lewis, were members of the Cambridge expedition to Vatnajokull, Iceland, in 1932. E. A. Shackleton, now Lord Shackleton, was a member, as surveyor, of the Oxford expedition to Sarawak in 1932. This expedition was made up of six Oxford men and three from Cambridge. Brian Roberts was also with the Cambridge expedition to Scoresby Sound, East Greenland, and A. Courtauld was a member of the Rasmussen Land Party. J. C. G. Sugden, an Oxford geographer, and father of David Sugden, now a member of the staff of the department of geography at Edinburgh, made two expeditions to Greenland in 1936 and 1938. Brian Roberts and A. Stephenson, who made his career after leaving Cambridge at Imperial College, were both members of the British Grahamland Expedition led by J. R. Rymill, 1934–7. John Wright, a surveyor by training and profession, was a member of the Cambridge visit to Ellesmere Land in 1938. The Imperial College visit to Jan Mayen included J. N. Jennings as glaciologist. W. G. V. Balchin, N. Pye and L. H. McCabe, who died in Hong Kong during the Japanese occupation, were members of a group that investigated nivation and corrie erosion in West Spitsbergen in 1938, and Balchin also published a paper (1941) on the magnificent series of raised beaches at Billefjord and Sassenfjord. A small Cambridge party (W. V. Lewis, J. N. Jennings, A. A. L. Caesar and M. Milne) visited Iceland (Vatnajokull) to make certain observations on cirque formation in 1933.

Since the war expeditions have multiplied and geographers, several of whom have become university teachers, have played a full part in them. In expeditions in which they have shared, geographers have nearly always been chosen for their expertise in some branch of physiography or as surveyors. Since 1945 many expeditions have gone forth for climbing in the Himalayas, Andes and other ranges; others have investigated human problems. The time will soon be ripe for a more comprehensive view of all the work done by young geographers on expeditions, and it will be found that they have added very considerably to our knowledge of many places and to the problems those places present.

REFERENCES

M. A. Arber (1940), 'The coastal landslips of south-east Devon', *Proceedings of the Geologists' Association*, 51, 257.

W. G. V. Balchin (1937), 'The erosion surfaces of North Cornwall', *Geographical Journal*, 90, 52–63.

(1941), 'The raised features of Billefjord and Sassenfjord, West Spitsbergen', *Geographical Journal*, 97, 364–76.

D. R. Brunsden and D. K. C. Jones (1976), 'The evolution of landslide slopes in Dorset', *Philosophical Transactions of the Royal Society*, A, 253, 605.

E. Derbyshire *et al.* (1979), 'Recent movement of the cliff at St Mary's Bay, Brixham, Devon', *Geographical Journal*, 145, 86–96.

A. R. Dwerryhouse and A. A. Miller (1930), 'The glaciation of the Clun Forest, Radnor Forest and some adjoining districts', *Quarterly Journal of the Geological Society*, 86, 96–127.

J. F. N. Green *et al.* (1934), 'The river Mole: its physiography and superficial deposits', *Proceedings of the Geologists' Association*, 45, 35–69.

J. F. N. Green (1941), 'The high platforms of East Devon', *Proceedings of the Geologists' Association*, 52, 36–52.

C. F. W. R. Gullick (1936), 'A physiographic survey of West Cornwall', *Transactions of the Royal Geological Society of Cornwall*, 16, 381–99.

R. W. G. Hingston (1930), 'The Oxford University expedition to British Guiana', *Geographical Journal*, 76, 1–24.

J. N. Hutchinson (1936), 'Coastal landslides in cliffs of Pleistocene deposits between Cromer and Overstrand, Norfolk, England', *Laurits Bjernon Memorial Volume*, Oslo.

J. N. Jennings (1952), 'The origin of the Broads', *Royal Geographical Society Research Memoir*, no. 2.

O. T. Jones and W. V. Lewis (1941), 'Water levels in Fowlmere and other Breckland meres', *Geographical Journal*, 97, 158–79.

R. O. Jones (1939), 'The evolution of the Neath-Tawe drainage system, South Wales', *Proceedings of the Geologists' Association*, 50, 530–66.

O. D. Kendall (1936), 'The coast of Somerset', *Proceedings of the Bristol Naturalists' Society*, 8, 186.

C. Kidson and A. P. Carr (1959), 'The movement of shingle over the sea bed close inshore', *Geographical Journal*, 125, 380–9.

C. A. M. King and F. A. Barnes (1964), 'Changes in the configuration of the inter-tidal beach zone of part of the Lincolnshire coast since 1951', *Zeitschrift für Geomorphologie*, 8, 105–26.

P. Lake (1934), 'The rivers of Wales and their connection with the Thames', *Science Progress*, 29, 25–40.

J. M. Lambert *et al.* (1960), 'The making of the Broads: a reconsideration of their origin in the light of new evidence', *Royal Geographical Society Research Memoir*, no. 3.

W. V. Lewis (1931), 'The effect of wave incidence on the configuration of a shingle beach', *Geographical Journal*, 78, 129–48.

(1932), 'The formation of Dungeness Foreland', *Geographical Journal*, 80, 309–24.

(1938a), 'A melt-water hypothesis of cirque formation', *Geological Magazine*, 75, 249–65.

(1938b), 'The evolution of shoreline curves', *Proceedings of the Geologists' Association*, 49, 107–27.

(1945), 'Nick points and the curve of water erosion', *Geological Magazine*, 82, 256–66.

W. V. Lewis and W. G. V. Balchin (1940), 'Past sea-levels at Dungeness', *Geographical Journal*, 96, 258–85.

D. L. Linton (1932), 'The origin of the Wessex rivers', *Scottish Geographical Magazine*, 48, 146–66.

(1933a), 'The origin of the Tweed drainage system', *Scottish Geographical Magazine*, 49, 162–75.

(1933b), 'The "Tinto Glacier" and some glacial features in Clydesdale', *Geological Magazine*, 70, 549–54.

(1934), 'On the former connection between the Clyde and the Tweed', *Scottish Geographical Magazine*, 50, 82–92.

(1940), 'Some aspects of the evolution of the rivers Earn and Tay', *Scottish Geographical Magazine*, 56, 1–11 and 69–79.

D. L. Linton and S. W. Wooldridge (1938), 'Influence of Pliocene transgression in the geomorphology of south-east England', *Journal of Geomorphology*, 1, 40–54.

G. Manley (1932), 'The weather of the High Pennines', *Durham University Journal*, 28, 31–2.

A. A. Miller (1935), 'The entrenched meanders of the Herefordshire Wye', *Geographical Journal*, 85, 160–78.

(1937), 'The 600-foot plateau in Pembrokeshire and Carmarthenshire', *Geographical Journal*, 90, 148–59.

(1938), 'Pre-glacial erosion surfaces around the Irish Sea basin', *Proceedings of the Yorkshire Geological Society*, 24, 31–59.

(1939), 'River development in southern Ireland', *Proceedings of the Royal Irish Academy*, 45B, 321–54.

J. A. Steers (1926), 'Orford Ness', *Proceedings of the Geologists' Association*, 37, 306.

(ed.) (1960), *Scolt Head Island* (2nd edn).

H. H. Swinnerton (1931), 'Post-glacial deposits on the Lincolnshire coast', *Quarterly Journal of the Geological Society*, 87, 360.

(1936), 'The physical history of east Lincolnshire', *Transactions of the Lincolnshire Naturalists' Union*, 31, 91–100.

W. W. Williams (1947), 'The determination of gradients on enemy-held beaches', *Geographical Journal*, 109, 76–93.

S. W. Wooldridge (1926), 'The structural evolution of the London Basin', *Proceedings of the Geologists' Association*, 37, 162–96.

(1927), 'The Pliocene history of the London Basin', *Proceedings of the Geologists' Association*, 38, 49–132.

(1932), 'The physiographic evolution of the London Basin', *Geography*, 17, 99–116.

S. W. Wooldridge and A. J. Bull (1925), 'The geomorphology of the Mole Gap', *Proceedings of the Geologists' Association*, 36, 1–10.

S. W. Wooldridge and J. F. Kirkaldy (1936), 'River profiles and denudation chronology in southern England', *Geological Magazine*, 73, 1–16.

S. W. Wooldridge and D. L. Linton (1939), *Structure, scenery and drainage in South-east England*. (*Transactions of the Institute of British Geographers*, 10.)

11 Geographers and geomorphology in Britain between the wars

D. R. STODDART*

Alfred Steers has documented the expansion of departments of geography, and of teaching in physical geography within them, before 1945. It is clear from his survey that in research, if not in teaching, 'physical geography' meant geomorphology: for while some attention was given to meteorology, climatology, and to some extent pedology and biogeography, it was on the level of elementary service courses for students rather than as a contribution to new knowledge.

What was the intellectual context within which these developments took place? To what extent were the academic achievements of the new physical geography constrained by the slow and scattered nature of its institutional development? What was the attitude of the geologists, throughout the nineteenth century the natural custodians of landform studies, to these activities, and how did the physical geographers respond?

The relationship of geographers with the geologists was a critical one, in various ways and on a variety of levels. When geography first became established at Oxford and Cambridge in the 1880s, geologists were less than enthusiastic: indeed D. W. Freshfield called them 'the most forward of the would-be "chuckers-out" of geography from the Hall of Education' (Freshfield 1886: 704). Geology itself was becoming increasingly specialized, and saw much of physical geography as elementary background material with which it was properly concerned, while the study of landforms, scarcely then dignified as a distinct and autonomous field of knowledge, was an indispensable adjunct to their own reconstructions of recent geological history. While in some universities geography gained its inde-

* David Ross Stoddart, O.B.E. (b. 15 November 1937) graduated in the Geographical Tripos of the University of Cambridge in 1959. Since 1962 he has been a member of the staff of the Department of Geography in Cambridge, and is a Fellow of Churchill College.

pendence from the start, in others it remained formally linked to geology, and it was to the geologists in these circumstances that teaching in physical geography often fell. Some of the geographers, thus confined to only a portion of their field of study, reacted by rejecting the need for physical geography altogether: at University College London, S. W. Wooldridge (1949: 13) tells us, Lyde dismissed geomorphology out of hand as 'mere morbid futility'.

Nevertheless, geomorphology has now become a field of knowledge so dominated in Britain by geographers that it is tempting to think that this has always been the case since the acceptance of geography in the universities. Indeed Herries Davies (1985: 388) has suggested that geomorphology was 'firmly renounced' by the geologists 'in the closing decades of the nineteenth century', and that geographers were its natural inheritors. The reality was, however, different, at least until the Second World War. The point is an important one, for the continuing role of the geologists constrained not simply the institutional development of geography but also the intellectual content of 'the whole great new science' of geomorphology, as Herbertson termed it in 1901 (Howarth 1951: 155).

This can readily be demonstrated by the prevailing text-books in this field, which had a strong geological bias at least until the Second World War. The geologist J. E. Marr (who began lecturing on the new subject of 'geo-morphology' in the department of geography at Cambridge in 1906) led the way with his *The Scientific Study of Scenery*. This was first published in 1900 and long remained the leading work in its field, reaching a ninth edition forty-three years later, long after Marr was dead. Its flavour is indicated by the subtitle of another book of his on the geology of the Lake District – 'and the scenery as influenced by geological structure' (Marr 1916). *The Building of the British Isles* ('a study in geographical evolution') by A. J. Jukes-Browne, which was first published in 1888 and went into its last edition thirty-four years later, was unashamedly lithological and stratigraphic, and made no concessions to new morphological ideas in any of its successive editions. On a regional level A. E. Trueman's *The Scenery of England and Wales* (1938), still in print in a revised and retitled form after nearly fifty years, was also written by a geologist, and indeed its highly successful later competitor, Dudley Stamp's *Britain's Structure and Scenery* (1946), though written by a geographer, reflected its author's initial training as a geologist.

The leading textbook of the time, however, replacing compendious physiographies such as Hugh Robert Mill's *Realm of Nature* (1892: last edition 1932), was Philip Lake's *Physical Geography*. This appeared in a first edition in 1915. It was revised after the Second World War by

J. A. Steers, G. Manley and W. V. Lewis, all in the Cambridge department, and was issued in its fourth edition forty years after its first publication. It was a lucid and extremely influential book at all levels, initially in the universities and ultimately in the schools, covering the whole field of physical geography; I remember well the pleasure I found in its direct simplicity of language and concept when a copy was presented to me at school. Lake himself, Reader in geography at Cambridge, had started life as a professional geologist, and indeed was even better known for his *Text-book of Geology*, written with R. H. Rastall. This was first published in 1910, and by its last edition in 1947, shortly before Lake's death, it had become *Lake and Rastall's Textbook of Geology*. Indeed, during Lake's time, the Cambridge geography department was housed in the Sedgwick Museum of Geology, and Lake was as much a figure in that department as in geography: and the same was true of the direction of his research. All of these books dominated university teaching in physical geography for an extraordinarily long period before the Second World War. They had, however, a pedagogic rather than a research function, and it was not until the 1930s that geographers themselves began to make more ambitious efforts to debate and contribute to physical geography on a research rather than an educational level: these works, with different aims, I will discuss later.

The story of the interplay between the traditional geological concern with landforms as representing one of many types of evidence for the reconstruction of earth history and the rather different orientation of the new geomorphology, developed in the United States in the last quarter of the nineteenth century, is fascinating and still largely unexplored. W. M. Davis, geomorphology's most articulate and relentless advocate, first came to Britain in 1894, a visit which resulted in one of his most brilliant and stimulating papers, 'The development of certain English rivers' (Davis 1895).[1] This paper not only defined for half a century both the research method and the preferred interpretation of the landforms of England, but also laid claim to interpret the geological development of the past few tens of millions of years from purely morphological evidence.

Davis visited Britain repeatedly, and it is indeed a remarkable fact that he directed some of his most powerful statements directly to a British audience. 'The geographical cycle' appeared in the *Geographical Journal* for 1899; 'The drainage of cuestas' (written following a visit to the Oxford area in 1898) in the *Proceedings of the Geologists' Association* in the same year; a version of 'The geographical cycle in an arid climate' and 'The sculpture of mountains by glaciers' in 1906, one in the *Geographical Journal*, the other in the *Scottish Geographical Magazine*; and 'Glacial erosion in North Wales' in the *Quarterly Journal of the Geological Society* in 1909. That same year he lectured twice to the Royal Geographical Society, on 'The systematic description of landforms' and on the lessons of the Grand Canyon (a lecture he repeated at Cambridge in 1913) (Davis 1909b, 1909c). But it is perhaps significant in terms of the reception of his ideas that during his 'geographical pilgrimage from Ireland to Italy' in 1911, he was accompanied in Britain by geologists, not geographers (by Marr in Snowdonia and by O. T. Jones in central Wales) (Davis 1911); indeed British geographers always thought of Davis as a geologist, not a geographer, in spite of his credentials.

The immediate reaction of geographers to his ideas was mixed, as can be seen from the discussion at the Royal Geographical Society in 1909 following his paper on landform description. H. J. Mackinder disliked the new terminology, and saw it as 'making geography into a merely supplementary chapter of geology'. A. J. Herbertson went along with him:

> I find that the more I work at geography, the less I use morphological classifica-
> tions and their terms. Whether I am dealing with land forms, or plants, or
> man, I find I have to give up the morphological terminology in favour of one
> that is rather descriptive of function, or, at any rate, expresses the character
> of the form in terms of the influence that the form exerts on the other geographi-
> cal features rather than on the history of the form – to deal with vegetations
> rather than floras, with economic or culture groups rather than with races.
> So that while, as a morphologist, I am extremely interested in all that Prof.
> Davis has said, as a geographer I sympathize with Mr Mackinder. I believe
> we must employ, for most geographical purposes, characteristic physiological
> descriptions rather than purely genetic morphological terms. (Davis 1909c p.
> 322)

Mill broadened the attack from one on terminology alone to one on Davis's entire method:

> I must say ... that the geographical cycle requires Prof. Davis to use it to
> perfection. It is a method that seems to me peculiarly dangerous when it is
> attempted with imperfect experience. Prof. Davis has so vast and clear a know-
> ledge of the surface forms of the Earth, and of the geological processes that
> have led to the existing scenery, that he can safely use methods which, in
> the hands of another, might lead to erroneous conclusions. His method appears
> to me as one more for the master than the student, and I am afraid that his
> disciples will run away from him, and apply it in a way that will cause him
> anxiety at first and horror afterwards. (Davis 1909c)

Davis was defended by the geologist Lamplugh, who pointed out that Davis was 'a geologist who has lent himself to the geographers for their benefit', an explanation which can scarcely have mollified the critics, and Davis defended himself with passion, conviction, and a degree of ruthlessness. The whole discussion reveals the tensions generated by the increasingly divergent aims of the two disciplines, and it reflects, too, a deep suspicion of theorizing that characterized a solidly empirical British geography for the entire first half of the century.

The power of Davis's methods was, nevertheless, remarkable, especially in the way that histories could be reconstructed from a combination of river patterns and profiles, geological structures, and land surface forms. But the geologists could claim that, leaving the terminology to one side, the method had a longer ancestry than was being allowed, certainly back to Jukes's classic paper of 1862 on the rivers of southern Ireland. And they were continuing the tradition independently of Davis and with little if any mention of his ideas: Strahan, Jukes-Browne, O. T. Jones, R. O. Jones and T. Neville George on the Welsh rivers, Wills on the Midlands, Cowper Reed, Hollingworth, King and others in the north, Henry Bury, J. F. N. Green, Kirkaldy, Bull and others in the south. The geographers contributed to the debate during the twenties and thirties but could not dominate it. Philip Lake (1900; 1934) had a long-standing interest in the Welsh rivers. Austin Miller, also a geologist by training, wrote on the southern Irish question in addition to his well-known paper on the Herefordshire Wye (Miller 1935; 1939a). R. F. E. W. Peel (1941) worked on the North Tyne. It was D. L. Linton, however, who in the thirties began to develop a new and distinctive, and specifically geographical, style of analysis of river development, quite different from that of the geologists, first in his paper on the Wessex rivers (1932), then in a long series on those of Scotland (1933; 1934; 1940).

None of these papers, however, asked fundamental questions about how rivers worked: such questions were not thought to be important. It was enough to 'sketch, or better still photograph' the landforms, as Wooldridge subsequently claimed (1958: 31), and to allow the results to speak for themselves: 'esoteric researches in fluid mechanics' were unlikely, he felt, to add much to the comprehension so derived of what he considered to be 'in essence' simple processes.

This emphasis on drainage patterns and their historical development contributed to and was readily reinforced by more comprehensive studies of regional denudation chronology. Here again the geologists were conspicuous, from the time of George Barrow's paper (1908) on Bodmin Moor onwards. It is interesting that their major contributions were mainly to the geomorphology of Highland Britain (Trotter on the Alston Block, Versey on Yorkshire, Hollingworth and McConnell on the north-west), though J. F. N. Green and A. E. Trueman published on the south. Several of these workers were intrigued not simply by the detail of local events but also by theoretical and technical issues (McConnell 1939a; Hollingworth 1938), a debate in which Austin Miller (1939b) joined.

It was, however, in regional studies of erosion history, mainly in southern Britain, that geographers made their most distinctive contribution. Several followed the lead of the geologists, even publishing in geological journals (Miller 1939c). But it was Wooldridge, himself a geologist by training, who turned from his initial petrographic studies of the Thames Basin to concern himself with details of surface form as indicative of history. After his first major paper in 1927 he combined with Linton, who in his Wessex work had analysed the relationship of drainage to structures, ultimately producing their synthesis *Structure, Surface and Drainage in South-east England* in 1939. Small (1980: 49) called this study 'the most persuasive and masterful piece of writing that British geomorphology had produced in half a century'. Though by the accident of its date of publication its impact was diminished by the Second World War (Brown and Waters 1974: 5), it nevertheless defined the nature of field research, the kinds of questions to be asked, and the techniques to be employed in the geomorphology of Britain for the next twenty years. Wooldridge especially, after his appointment to the chair at King's College London in 1947, directed a cohort of graduate students and colleagues to work over this ground and extend it. Part at least of the appeal of what was, on one level, a regional study deeply rooted in the intricacies of local detail, was the breadth and generality of its implications. Indeed the concept of a eustatic Pliocene transgression (Wooldridge 1928; Wooldridge and Linton 1938) supplied an immediate tool for the interpretation of landforms throughout the rest of the country (e.g. Miller 1937; Balchin 1937; Green 1941).

There were, of course, many other studies in local geomorphology during the twenties and thirties, especially in southern England (where most of the university geographers were located) – on the drifts and the course of the Thames, for example (clearly under Wooldridge's influence: Wooldridge 1938), and on dry valleys and other features of the English Chalk, especially in the Weald (Fagg 1923; Bull 1936), and all of them made some contribution to the grand design of regional historical synthesis.

There is a danger, I feel, of too readily retrospectively classifying all of this work as 'Davisian'. Indeed Brown and Waters (1974: 3) comment that 'the impact of his (Davis's) ideas seems to have been minimal', at least up to the end of the First World War. One reason why Davis's influence is often seen as greater than it was, apart from the obvious similarity of ideas in historical reconstruction, was the constant public defence of Davis and his ideas by Wooldridge (1955; 1958) in his later years. It is true that he and Morgan made Davis's cycle of erosion the 'central theme and method' of their *Physical Basis of Geography: an outline of geomorphology*, published in 1937.[2] But they also recalled that 'in our view this very powerful and flexible method of study has never had justice done to it in Britain' (1937: viii). Perhaps this was because, to geologists like Barrow and Bury, it was unnecessary: they were doing it already, and their empirical results had little need of either the concepts or the lexicon which Davis provided. Nor did Wooldridge and his colleagues introduce new techniques unknown to the geologists. True, he mapped the Pliocene bench, but benches were being mapped by J. F. N. Green, R. B. McConnell, and virtually everyone else at that time. The fact is that Wooldridge's own methods were mainly those of the field geologist, based on 'detailed knowledge of the ground ... It was', says K. M. Clayton (1980: 9–10), 'the proud aim of Wooldridge and Linton that they should know all the ground, and that the name of a village would be enough to recall a site, a pattern of relief and geology, that was a piece of the jigsaw they had set out to reconstruct ... They had both walked across every parish in south-east England.' And it was this level of detailed knowledge which made their conclusions so difficult to controvert.

It is remarkable that, with one outstanding exception, the work so far mentioned completed the geographers' contribution to geomorphology. Glacial studies had long been the preserve of the geologist. W. B. Wright lucidly summarized their results in *The Quaternary Ice Age* (1914; 2nd edn 1937), a study still being recommended to students at Cambridge half a century after its first publication. In spite of Davis's (1909a) persuasive diagrams of Snowdonia and of W. H. Hobbs's (1910) exegesis of 'The cycle of mountain glaciation' in the *Geographical Journal* (35: 146–63; 268–84) geographers paid no attention to research on glacial geomorphology. The literature before 1939 was dominated by geologists such as Harmer, Kendall, Dwerryhouse, Raistrick, Trotter and Hollingworth, and not surprisingly consisted mainly of painstaking reconstructions of the course of events from the morphostratigraphy of depositional features. To the extent that wider questions were considered, as over the problem of glacial protection, the debate was by geologists (e.g. Garwood 1910). It was not until the eve of the Second World War that W. Vaughan Lewis at Cambridge changed the course of glacial studies by asking questions not of history but of process: and to do so he had to go to where the ice was, not where it once had been (Lewis 1938b; 1939; 1940). This in itself was a major innovation; but then the war interrupted his work virtually for a decade, and the development of his ideas belongs largely to the late forties and the fifties.

Periglaciation had scarcely been discovered, of course, in spite of some interest in the stone stripes of the Lake District and in the characteristics of 'head': the implications of the discovery by T. T. Paterson (1940), freshly back from Baffin Bay, of ice wedges in a quarry wall behind the Traveller's Rest public house in Cambridge were again postponed to later times by the war. Wooldridge (1958: 30) in any case thought periglaciation a 'craze', and this was a powerful disincentive for anyone to do anything about it.

It is, however, an extraordinary fact that throughout the period under discussion there was no shortage of observations on glacial and related phenomena, mostly from the Arctic, but also from the great mountain ranges, notably the Himalaya, published mainly in the *Geographical Journal*. These were very far from being simply the expedition narratives of climbers and explorers. The studies by N. E. Odell (1933; 1937), himself an Everest mountaineer of great distinction, on the glaciated mountains of Labrador and Greenland, and of McCabe on Spitsbergen (1939), suffice to make the point, which could be repeatedly duplicated from the literature of the twenties and thirties. The same applies to research on the tropical deserts. Throughout the thirties the *Geographical Journal* carried a remarkable series of papers on the Libyan Desert (Beadnell 1910, 1934; Ball 1927; Sandford 1933; Kadar 1934; Kennedy-Shaw 1936), in which, in retrospect, the accounts of R. A. Bagnold's expeditions (1931; 1933), culminating in the multi-disciplinary Gilf Kebir project (Bagnold *et al.* 1939), had greatest potential for future systematic study – achieved in Bagnold's own *The Physics of Blown Sand and Desert Dunes* (1941). Vaughan Cornish had long since written on these topics (1897; 1914), and some of Davis's last and longest papers were on desert erosion (Davis 1930; 1938). His paper on the arid cycle had itself been published in this country. Yet British geographers made no contribution in this field: indeed until Peel joined the Gilf Kebir expedition they made no contribution at all to arid geomorphology. As for the humid tropics and the savanna lands, they might not have existed. Think of the German researchers at the same period! – Passarge in the Kalahari, Walther in Sinai, Obst and Bornhardt in East Africa, Kaiser and Waibel in south-west Africa, Freise in Brazil, Jaeger, Jessen, Credner, Krebs, Thorbecke, and especially Sapper, around the world in the humid tropics. And the list could readily be duplicated with the French, in Indo-China and the Sahara.

What is surprising is not only that British geographers did not make a contribution in these fields, but that they quite failed to incorporate the results published in the *Geographical Journal* into the corpus of academic knowledge (see, for example, the material cited in Chapters 20 and 22 of Wooldridge and Morgan's *The Physical Basis of Geography* (1937), and contrast Cotton's *Climatic Accidents in Landscape-making* (1942)).[3] There are reasons for the lack of active participation in such studies which I shall return to later, but it is worth considering why the work that was done by others was so resolutely ignored by the geographers in the universities. Undoubtedly this results from the character and interests of dominant figures like Wooldridge, inextricably committed to the inch-by-inch perusal of the English countryside. It was in the Preface to *The Physical Basis of Geography* (1937: x) that he quoted with evident relish and approval Proverbs 17: 24, 'The eyes of the fool are in the ends of the earth',[4] and stated baldly that 'geomorphology must begin at home if the student is to cultivate the "eye for country" which alone can make him the master of his medium'. Elsewhere he made clear his view that studies of Somerset should take priority over those of Somalia, and made the bizarre assertion that the geography of the former was of a higher intellectual order than that of the latter (Wooldridge 1952: 7). It would not surprise me to learn that Wooldridge himself never left the shores of Britain. Certainly, and unusually, he kept aloof from the European excursions of the Le Play Society. As a blunt Congregationalist he had nothing in common, socially, emotionally, academically, or intellectually, with the explorers of the earth's wild places in the Royal Geographical Society: the distance between the Strand and Kensington Gore was not to be measured simply in miles. And neither the Secretary of the Royal Geographical Society, Arthur Hinks, nor Wooldridge himself were temperamentally the kind of men to make allowances.

One cannot, therefore, be surprised that it was Wooldridge who took the lead in taking the red-brick academic geographers out of the uncongenial halls of the Royal Geographical Society, past Gino Watkins's kayak, Douglas Mawson's sledge, and all the other memorabilia of great achievement at the ends of the earth, into the homeless Institute of British Geographers when the latter was founded in 1933 (Steel 1984). The narrowing of world horizons that such attitudes involved has scarcely yet been gauged; but it is ironic that it was in large degree Wooldridge, so concerned about taking the *ge-* out of geography (Wooldridge 1949), who made so little effort to incorporate the *geo-*.

The consequences of this perverse parochialism were not simply in the loss of huge areas of experience, but in a restriction of intellectual horizons. Consider the overarching ideas of geomorphology in the first half of this century. The cycle of erosion – yes; but its influence in research was surely exaggerated, and its value mainly as a pedagogic device, useful more in structuring text-books than in generating new knowledge. What of the rest?

First, the problems of tectonics and of the mobility or permanence of the continents. This was, after all, the time of fundamental reappraisals of Alpine (and Himalayan) structures, as well as of the successive editions of Alfred Wegener's *Die Entstehung der Kontinente und Ozeane* (1914; later editions 1920, 1922, 1929; first English translation 1924). These were fully reviewed and discussed, both in Steers's *The Unstable Earth* (1932) and in Wooldridge and Morgan's *The Physical Basis of Geography* (1937) (in the latter case much more sympathetically than one might have expected given the attitudes of Wooldridge's later years[5]), but without making much connection, if any, to the problems of landforms and their evolution.

Second, that of the legacy of Pleistocene events. Here interpretation was overshadowed by the massive achievement of Penck and Brückner's *Die Alpen im Eiszeitalter* (1909). Partly because of the extensiveness of the later glaciations in Britain, partly because of the inherent ambiguity of the drifts (which still is unresolved), British workers developed less ambitious aims: their results are primarily of local rather than theoretical significance.

Third, and reinforcing the study of glacial chronology, was the synthesis of episodic eustatic sea-level change coupled with that of glacial advance and retreat, initially derived from the Mediterranean sequences of Depéret. In Britain attention was directed on the one hand to the late-Tertiary record of benches and terraces, exemplified by *Structure, Surface and Drainage in South-east England*, and on the other to the intricacies of Pleistocene coastal formations. It is remarkable (and to some, at the time, galling) that it was left to a Frenchman from Strasbourg, Henri Baulig, to instruct the members of the Institute of British Geographers on 'The changing sea-level' in 1935 (Baulig 1935).

Fourth, the reality of continuing environmental change, especially in the arid and semi-arid lands. Ellsworth Huntington had dramatically drawn attention to this in *The Pulse of Asia* and later works (Huntington 1907; 1914), and British geologists and archaeologists such as K. S. Sandford and W. J. Arkell were demonstrating its reality in Egypt and the Middle East. Yet because British physical geographers drew back from active involvement in the tropics, this key to earth history and process remained unturned in the period under review. Provocative works of planetary synthesis such as R. A. Daly's *The Changing World of the Ice Age* (1934) were received with polite attention but scarcely enthusiasm.

Notice that I do not include in this list of dominant ideas what would now be called theoretical geomorphology. Hardly anyone was interested in questions of morphometry or process. Harold Jeffreys (1918) wrote on 'Problems of denudation' and Philip Lake (1928) 'On hillslopes': and that was that. With hindsight one might well ask about the impact of Walther Penck's *Die morphologische Analyse* (1924), regarded in the fifties and sixties by the text-book writers as the great challenge and alternative to Davis's cyclic scheme. 'J.A.S.', reviewing it in the *Geographical Journal* in 1926, found it 'interesting', its argument 'intricate', and the book as a whole 'extremely difficult reading'. It made no impact on geomorphology generally until the symposium on Penck at the Association of American Geographers, organized by von Engeln in 1940, and its subsequent discussion in von Engeln's *Geomorphology* (1942) and in Cotton's *Landscape as Developed by the Processes of Normal Erosion* (1941). Penck's book itself was not translated into English until 1953, and the time for its impact to be made had then passed.

There is, nevertheless, one area of geomorphology in Britain in the thirties to which the generalizations I have made did not apply: and this was in the study of coasts. In the first place, geographers dominated the literature on coastal studies from an early date (though geologists such as Green and George were active in studying the Pleistocene legacy of raised beaches and coastal terraces). Second, while of course there were many detailed local studies of particular areas and features (e.g. Ogilvie 1914, 1923 in Scotland; and Steers 1926b, 1927, 1934a, 1937a and 1939, in precursor studies to *The Coastline of England and Wales* (1946)), there was also an immediate interest in processes and resultant forms. Perhaps this was inherent in the subject matter itself, given the rapidity of adjustment of form to changing processes on beaches. Lewis at Cambridge led the way with original and innovative papers throughout the 1930s (Lewis 1931; 1932; 1938a). Moreover, coastal studies, almost uniquely, were pursued overseas, first by Steers's participation in the Great Barrier Reef Expedition in 1928–9, then by his geographical expedition to the Great Barrier Reefs in 1936, and later by his work on the Cambridge Expedition to Jamaica in 1939. And finally, and again perhaps implicit in the subject matter, much of this work was collaborative and interdisciplinary, as shown by Steers's book on the dunes and salt marshes of Scolt Head Island (1934b) and by his association both there and in Jamaica with the Cambridge botanist and ecologist V. J. Chapman.

The example of coastal geomorphology, and the role played by the Cambridge department in its development, prompts me to ask why other branches of investigation were not pursued with similar intellectual curiosity. It would be too easy, perhaps, to suggest that times were difficult and money hard to find, and that this was why so many stayed at home. Such circumstances did not inhibit Haddon and Seligman and a host of 'anthropo-geographers' and anthropologists from immersing themselves in local cultures in New Guinea, West Africa and around the world. Indeed the editor of this volume himself showed what could be done when he walked with porters through the hinterland of Sierra Leone as a postgraduate student in 1938.

I think the answer is more straightforward: there were simply too few people involved to do everything that needed to be done. British geomorphology (and *a fortiori* the rest of physical geography) throughout the thirties was represented by Wooldridge and Linton, Steers and Lewis, Austin Miller – and hardly anybody else. Not surprisingly they worked on what interested them, and inevitably it added up to a patchy and almost idiosyncratic agenda for research. Why the numbers were so small, and why, for example, the development of geography in British universities lagged so far behind that in France and Germany, both in timing and in scale, raises questions of national priorities and educational policies which cannot be debated here.

The situation, however, had two important implications, and we may borrow concepts from the natural sciences to describe them. The first is the operation of a 'founder principle'. Wooldridge, Steers and Linton all became holders of chairs and senior figures in the geographical establishment: 'dominating figures', in Small's words, 'influencing the ideas and methods of many younger workers who came under their sway' (Small 1980: 49). King's College under Wooldridge produced a long line of like-minded students, who, under his patronage, themselves came to occupy the chairs of an expanding university system. And from the quite different corridors of Cambridge, and specifically of St Catharine's College, Steers did the same, and on an even wider scale. The dominant figures of the next generation of British geomorphologists – J. N. Jennings, M. M. Sweeting, C. A. M. King, A. T. Grove, E. H. Brown, B. W. Sparks – owed their careers and their interests to the intellectual atmosphere and opportunities created in Cambridge and London by the small group of men I have described. Perhaps by temperament as well as by location, Linton never exercised the same kind of influence from Sheffield and Birmingham, and Lewis, tragically, died too soon (but not before launching a cohort of glaciologists on the world scene).

168 D. R. Stoddart

The second implication is that the numbers were so small and their locations so diverse, that the 'critical mass' for interaction was missing. To a large degree self-excluded from Kensington Gore, they had nowhere else to meet to discuss common interests other than the annual meeting of the British Association for the Advancement of Science – and of course the meetings of the geological societies, where they were automatically outnumbered and on the defensive. It is in this perspective that the foundation of the Institute of British Geographers can be seen in its true importance: it was an institutional recognition of the fact that for the first time there were enough professional academics (and the entrance qualifications were set deliberately high) to interact with each other, to exchange ideas, and ultimately to collaborate in research. The lone individuals, ploughing (as the accounts in this book show) a generally lonely and often thankless furrow, beset by the demands of teaching and examining, generally ill-paid, were soon to give way to research schools and fully fledged departments.

But in large degree the emphases and traditions of this later age derived directly from those, so hardly won, of this earlier time. One looks back to it and to its leading figures with affection and respect, and – be it said – with a degree of nostalgia for a world that was doubtless harder in many respects than our own, but was undoubtedly a great deal simpler and less regimented, and which has gone for ever.[6]

NOTES

[1] It is extraordinary that this classic paper is mistitled by everyone who cites it, not only in the memorial volume for David Linton (Brown and Waters 1974) but also in the volume commemorating the fortieth anniversary of the publication of Wooldridge and Linton's *Structure, Surface and Drainage in South-east England* (Jones 1980).
[2] This book remained the only textbook on geomorphology in common use in Britain until outdated by Thornbury's *Principles of Geomorphology* in the mid-1950s. Its intellectual roots in geology are very clearly demonstrated by its citation structure, which is quite unlike that of other books on the subject. Most of the references cited in it date, as would be expected, from the few years prior to its publication, but there is a distinct and substantial secondary mode of material published between 1895 and 1915: indeed 40 per cent of all the references date from before 1920, and 16 per cent from before 1900. The reason is that for Wooldridge's type of geomorphology a fundamental source was the regional memoirs of the Geological Survey, which in their massive local detail supplied a factual validity independent of their age. In most sciences, of course, and in geomorphology subsequently, it is the development of ideas that is important, and the content and accelerating advancement

of knowledge is reflected in citation structures. *The Physical Basis of Geography* was still in use as an unrevised reprint when I was an undergraduate in the late 1950s. By that time more than half the references in it were more than thirty years old, and a quarter more than half a century: this hardly made for intellectual excitement, irrespective of the ideas the book discussed. It is interesting to compare Wooldridge and Morgan's book with Steers's *The Unstable Earth*, first published in 1932. Steers introduced a guide to sources and further reading in the second edition in 1937. Almost two-thirds of the citations are dated 1935 or later, and none is earlier than 1931. This immediacy in the referencing reflected the rate of scientific advance quite as forcibly as the two dominant themes he singled out for discussion – that of the evolution of continents and oceans, and of changes of sea-level in Pleistocene and Recent times. Not surprisingly the book retained its relevance until its final reprinting in 1955, long after Wooldridge and Morgan's had become conceptually obsolete.

3 Perhaps the only paper in this genre which made any real impact on the collective geomorphic imagination – not without cause – was that by L. R. Wager (1937), in which he demonstrated the spectacular antecedence of the course of the Arun as it cut through the Himalaya from the Tibetan Plateau. Wager, of course, was a geologist, and a hard-rock one at that.

4 The text as quoted differs from that both of the King James Bible and the Revised Standard Version. The form given in the *New English Bible*, which one would not normally consult ('a stupid man's eyes are roving everywhere'), invites a quite different interpretation from that given by Wooldridge.

5 I recall meeting Wooldridge on Dartmoor when I was a schoolboy, in (I think) 1954. I asked him his views on continental drift. The result was, to say the least, alarming; in fact I feared I had provoked an apoplexy. This was, of course, after his stroke, and he was no longer at the height of his powers.

6 It is curious that there have been so few studies of the development of British geomorphology during the twenties and thirties. Linton (1969) discussed the history of work in lowland England, and Wooldridge (1951) progress in more general terms. Brown and Waters (1974) pass rapidly over the period before 1939. Dury's account of geomorphology over the last fifty years, prepared for the fiftieth anniversary of the Institute of British Geographers (Dury 1983), is hardly without bias: one-third of all the works he cites as being significant in this period are by himself.

REFERENCES

W. J. Arkell (1943), 'The Pleistocene rocks at Trebetherick Point, north Cornwall: their interpretation and correlation', *Proceedings of the Geologists' Association*, 54: 141–70.

R. A. Bagnold (1931), 'Journeys in the Libyan Desert', *Geographical Journal*, 78: 13–39, 524–35.

(1933), 'A further journey in the Libyan Desert', *Geographical Journal*, 82: 103–29, 403–4.

(1941), *The Physics of Blown Sand and Desert Dunes*, xx, 265 pp.

et al. (1939), 'An expedition to the Gilf Kebir', *Geographical Journal*, 93: 281–313.

W. G. V. Balchin (1937), 'The erosion surfaces of north Cornwall', *Geographical Journal*, 90: 52–63.

J. Ball (1927), 'Problems of the Libyan Desert', *Geographical Journal*, 70: 21–38, 105–128, 209–24.

G. Barrow (1908), 'The high level platforms of Bodmin Moor and their relation to the deposits of stream-tin and wolfram'. *Quarterly Journal of the Geological Society of London*, 64: 384–400.

H. Baulig (1935), 'The changing sea-level', *Transactions and Papers of the Institute of British Geographers*, 3: 1–46.

H. J. L. Beadnell (1910), 'The sand dunes of the Libyan Desert', *Geographical Journal*, 35: 379–95.

(1934), 'Libyan desert dunes', *Geographical Journal*, 84: 337–40.

E. H. Brown and R. S. Waters (1974), *Geomorphology in the United Kingdom since the First World War (Special Publications of the Institute of British Geographers)*, 7: 3–9.

(eds) (1974), *Progress in Geomorphology: Papers in honour of David L. Linton. (Special Publications of the Institute of British Geographers*, 7: 1–251.)

A. J. Bull (1936), 'Studies in the geomorphology of the South Downs', *Proceedings of the Geologists' Association*, 47: 99–129.

(1942), 'Pleistocene chronology', *Proceedings of the Geologists' Association*, 53: 1–45.

H. Bury (1910), 'The denudation of the western end of the Weald', *Quarterly Journal of the Geological Society of London*, 66: 640–92.

(1926), 'The rivers of the Hampshire Basin', *Proceedings of the Hampshire Field Club*, 10: 1–12.

K. M. Clayton (1980), 'The historical context of *Structure, Surface and Drainage in South-east England*', in D. K. C. Jones (ed.), *The Shaping of Southern England (Special Publications of the Institute of British Geographers*, 11: 1–12.)

V. Cornish (1897), 'On the formation of sand dunes', *Geographical Journal*, 9: 278–309.

(1914), *Waves of sand and snow.* 383 pp.

C. A. Cotton (1941), *Landscape as Developed by the Processes of Normal Erosion.* xviii, 302 pp.

(1942), *Climatic Accidents in Landscape-making.* xx, 343 pp.

R. A. Daly (1934), *The Changing World of the Ice Age.* xix, 271 pp.

W. M. Davis (1895), 'The development of certain English rivers', *Geographical Journal*, 5: 127–46.

(1899a), 'The geographical cycle', *Geographical Journal*, 14: 481–504.

(1899b), 'The drainage of cuestas', *Proceedings of the Geologists' Association,* 16: 75–93.

(1906a), 'The geographical cycle in an arid climate', *Geographical Journal,* 27: 70–3 (previously published in full in *Journal of Geology,* 13 (1905): 381–407).

(1906b), 'The sculpture of mountains by glaciers', *Scottish Geographical Magazine,* 22: 76–89.

(1909a), 'Glacial erosion in north Wales', *Quarterly Journal of the Geological Society of London,* 65: 281–350.

(1909b), 'The Colorado Canyon: some of its lessons', *Geographical Journal,* 33: 535–40.

(1909c), 'The systematic description of land forms', *Geographical Journal,* 34: 300–18; discussion, 318–26.

(1911), 'A geographical pilgrimage from Ireland to Italy', *Annals of the Association of American Geographers,* 2: 73–100.

(1930), 'Rock floors in arid and humid climates', *Journal of Geology,* 38: 1–27, 136–8.

(1938), 'Sheetfloods and streamfloods', *Bulletin of the Geological Society of America,* 49: 1337–416.

H. G. Dines, S. E. Hollingworth, W. Edwards, S. Buchan and F. B. A. Welch (1940), 'The mapping of head deposits', *Geological Magazine,* 77: 198–226.

G. H. Dury (1983), 'Geography and geomorphology: the last fifty years', *Transactions of the Institute of British Geographers,* N.S., 8: 90–9.

O. D. von Englen (1942), *Geomorphology,* ix, 655 pp.

C. C. Fagg (1923), 'The recession of the Chalk escarpment and the development of Chalk valleys', *Proceedings and Transactions of the Croydon Natural History and Scientific Society,* 9: 93–112.

D. W. Freshfield (1886), 'The place of geography in education', *Proceedings of the Royal Geographical Society,* N.S., 8: 698–714.

E. J. Garwood (1910), 'Features of Alpine scenery due to glacial protection', *Geographical Journal,* 36: 310–39.

T. N. George (1932), 'The Quaternary beaches of Gower', *Proceedings of the Geologists' Association,* 43: 291–324.

(1942), 'The development of the Towy and upper Usk drainage pattern', *Quarterly Journal of the Geological Society of London,* 98: 89–137.

J. F. N. Green (1934), 'The River Mole: its physiography and superficial deposits', *Proceedings of the Geologists' Association,* 45: 35–69.

(1941), 'The high platforms of East Devon', *Proceedings of the Geologists' Association,* 52: 36–52.

(1943), 'The age of the raised beaches of south Devon', *Proceedings of the Geologists' Association,* 54: 129–40.

F. W. Harmer (1928), 'The distribution of erratics and drift', *Proceedings of the Yorkshire Geological Society,* 21: 79–150.

G. L. Herries Davies (1985), 'James Hutton and the study of landforms', *Progress in Physical Geography*, 9: 382–9.

W. H. Hobbs (1910), 'The cycle of mountain glaciation', *Geographical Journal*, 35: 146–63, 268–84.

S. E. Hollingworth, (1929), 'The evolution of the Eden drainage in the south and west', *Proceedings of the Geologists' Association*, 40: 115–38.

(1931), 'The glaciation of western Edenside and adjoining areas, and the drumlins of Edenside and the Solway', *Quarterly Journal of the Geological Society of London*, 87: 281–359.

(1936), 'High-level erosional platforms in Cumberland and Furness', *Proceedings of the Yorkshire Geological Society*, 23: 159–77.

(1938), 'The recognition and correlation of high level erosion surfaces in Britain: a statistical study', *Quarterly Journal of the Geological Society of London*, 94: 55–84.

O. J. R. Howarth (1951), 'The centenary of Section E (Geography) in the British Association', *Scottish Geographical Magazine*, 67: 145–60.

E. Huntington (1907), *The Pulse of Asia*. xxi, 415 pp.

(1914), *The Climatic Factor as Illustrated in Arid America (Publications of the Carnegie Institution of Washington)*, 192: 1–341.

H. Jeffreys (1918), 'Problems of denudation', *Philosophical Magazine*, 36: 179–90.

O. T. Jones (1924), 'The upper Towy drainage system', *Quarterly Journal of the Geological Society of London*, 80: 568–609.

R. O. Jones (1931), 'The development of the Tawe drainage', *Proceedings of the Geologists' Association*, 42: 305–21.

(1939), 'The evolution of the Neath-Tawe drainage system, South Wales', *Proceedings of the Geologists' Association*, 50: 530–66.

J. B. Jukes (1862), 'On the mode of formation of some of the river-valleys in the south of Ireland', *Quarterly Journal of the Geological Society of London*, 18: 378–403.

A. J. Jukes-Browne (1888), *The Building of the British Isles: a study in geographical evolution* (London, G. Bell), x, 343 pp. (4th edn, London, E. Stanford 1922). xv, 470 pp.)

(1904), 'The valley of the Teign', *Quarterly Journal of the Geological Society of London*, 60: 319–34.

L. Kadar (1934), 'A study of the sand sea in the Libyan Desert', *Geographical Journal*, 83: 470–8.

P. F. Kendall (1902), 'A system of glacier lakes in the Cleveland Hills', *Quarterly Journal of the Geological Society of London*, 58: 471–571.

W. B. Kennedy-Shaw (1936), 'An expedition to the south Libyan Desert', *Geographical Journal*, 87: 193–221.

W. B. R. King (1935), 'The upper Wensleydale river system', *Proceedings of the Yorkshire Geological Society*, 63: 10–24.

J. F. Kirkaldy and A. J. Bull (1940), 'The geomorphology of the rivers of the southern Weald', *Proceedings of the Geologists' Association*, 51: 115–50.

P. Lake (1900), 'Bala Lake and the river system of North Wales', *Geological Magazine* (4) 7: 204–15, 241–5.

  (1915), *Physical Geography*. xx, 324 pp. (2nd edn revised by J. A. Steers, G. Manley and W. V. Lewis, edited by J. A. Steers, 1949, xxviii, 410 pp.; 3rd edn likewise, 1955, xxviii, 424 pp.)

  (1928), 'On hill-slopes', *Geological Magazine*, 65: 108–16.

  (1934), 'The rivers of Wales and their connection with the Thames', *Science Progress*, 29: 25–40.

P. Lake and R. H. Rastall (1910), *A Text-book of Geology*. xvi, 494 pp. (6th edn as *Lake and Rastall's Textbook of Geology*, revised by R. H. Rastall, 1947, viii, 490 pp.)

W. V. Lewis (1931), 'The effect of wave incidence on the configuration of a shingle beach', *Geographical Journal*, 78: 129–48.

  (1932), 'The formation of Dungeness Foreland', *Geographical Journal*, 80: 309–24.

  (1938a), 'The evolution of shoreline curves', *Proceedings of the Geologists' Association*, 49: 107–27.

  (1938b), 'A melt-water hypothesis of cirque formation', *Geological Magazine*, 75: 249–65.

  (1939), 'Snow patch erosion in Iceland', *Geographical Journal*, 94: 153–61.

  (1940), 'The function of meltwater in cirque formation', *Geographical Review*, 30: 64–83.

D. L. Linton (1932), 'The origin of the Wessex rivers', *Scottish Geographical Magazine*, 48: 149–66.

  (1933), 'The origin of the Tweed drainage system', *Scottish Geographical Magazine*, 49: 162–74.

  (1934), 'On the former connection between the Clyde and the Tweed', *Scottish Geographical Magazine*, 50: 82–92.

  (1940), 'Some aspects of the evolution of the rivers Earn and Tay', *Scottish Geographical Magazine*, 56: 1–11, 69–79.

  (1969), 'The formative years in geographical research in south-east England', *Area*, 1 (2): 1–8.

J. E. Marr (1900), *The Scientific Study of Scenery*. ix, 368 pp. (9th edn, 1943, ix, 372 pp.)

  (1916), *Geology of the Lake District and the Scenery as influenced by Geological Structure*. xii, 220 pp.

L. H. McCabe (1939), 'Nivation and corrie erosion in West Spitsbergen', *Geographical Journal*, 94: 447–65.

R. B. McConnell (1939a), 'Residual erosion surfaces in mountain ranges', *Proceedings of the Yorkshire Geological Society*, 24: 76–98.

  (1939b), 'The relic surfaces of the Howgill Fells', *Proceedings of the Yorkshire Geological Society*, 24: 152–64.

174   D. R. Stoddart

H. R. Mill (1892), *The Realm of Nature: an outline of physiography*. xii, 369 pp. (3rd edn, 1924, xii, 404 pp.; final reprint 1932.)

A. A. Miller (1935), 'The entrenched meanders of the Herefordshire Wye', *Geographical Journal*, 85: 160–78.

(1937), 'The 600-foot plateau in Pembrokeshire and Carmarthenshire', *Geographical Journal*, 90: 148–59.

(1939a), 'River development in southern Ireland', *Transactions of the Royal Irish Academy*, 45B: 321–54.

(1939b), 'Attainable standards of accuracy in the determination of pre-glacial sea levels by physiographic methods', *Journal of Geomorphology*, 2: 95–115.

(1939c), 'Pre-glacial erosion surfaces around the Irish Sea basin', *Proceedings of the Yorkshire Geological Society*, 24: 31–59.

N. E. Odell (1933), 'The mountains of northern Labrador', *Geographical Journal*, 82: 193–210, 315–25.

(1937), 'The glaciers and morphology of the Franz Josef region of north-east Greenland', *Geographical Journal*, 90: 111–25, 233–58.

A. G. Ogilvie (1914), 'The physical geography of the entrance to Inverness Forth', *Scottish Geographical Magazine*, 30: 21–35.

(1923), 'The physiography of the Moray Firth coast', *Transactions of the Royal Society of Edinburgh*, 53: 377–404.

T. T. Paterson (1940), 'The effects of frost action and solifluction around Baffin Bay and in the Cambridge district', *Quarterly Journal of the Geological Society of London*, 96: 99–130.

R. F. E. W. Peel (1941), 'The North Tyne valley', *Geographical Journal*, 98: 5–19.

A. Penck and E. Brückner (1909), *Die Alpen im Eiszeitalter*. 3 vols, x, 1199 pp.

A. Raistrick (1926), 'Glaciation of Wensleydale, Swaledale, and adjoining parts of the Pennines', *Proceedings of the Yorkshire Geological Society*, 20: 366–411.

F. R. C. Reed (1901), *The Geological History of the Rivers of East Yorkshire, being the Sedgwick Prize Essay for 1900*. vi, 103 pp.

K. S. Sandford (1933), 'Past climate and early man in the southern Libyan Desert', *Geographical Journal*, 82: 219–22.

R. J. Small (1980), 'The Tertiary geomorphological evolution of south-east England: an alternative interpretation', in D. K. C. Jones (ed.), *The Shaping of Southern England (Special Publications of the Institute of British Geographers)*, 11: 49–70.

M. A. Spender (1930), 'Island reefs of the Queensland coast', *Geographical Journal*, 76: 193–214, 273–97.

L. D. Stamp (1946), *Britain's Structure and Scenery*. 255 pp.

R. W. Steel (1984), *The Institute of British Geographers: the first fifty years*.

J. A. Steers (1926a), Review of W. Penck's *Die morphologische Analyse: Eine Kapitel der physikalischen Geologie* (1924), *Geographical Journal*, 67: 272–3.

(1926b), 'Orford Ness: a study in coastal physiography', *Proceedings of the Geologists' Association*, 37: 306–25.

(1927), 'The East Anglian coast', *Geographical Journal*, 69: 24–48.

(1929), 'The Queensland coast and the Great Barrier Reefs', *Geographical Journal*, 74: 232–57, 341–70.

(1932), *The Unstable Earth: some recent views in geomorphology*. xiii, 341 pp. (5th edn, 1950, xv, 345 pp.; final reprint 1955.)

(1934), 'Scolt Head Island', *Geographical Journal*, 83: 479–502.

(ed.) (1934b), *Scolt Head Island, the Story of its Origin: the plant and animal life of the dunes and marshes*. xvi, 234 pp.

(1937a), 'The Culbin Sands and Burghead Bay', *Geographical Journal*, 90: 498–528.

(1937b), 'The coral islands and associated features of the Great Barrier Reefs', *Geographical Journal*, 89: 1–28, 119–46.

(1939), 'Sand and shingle formations in Cardigan Bay', *Geographical Journal*, 94, 209–27.

(1940a), 'Coral cays of Jamaica', *Geographical Journal*, 95: 30–42.

(1940b), 'The cays and the Palisadoes, Port Royal, Jamaica', *Geographical Review*, 30: 279–96.

(1946), *The Coastline of England and Wales*. xix, 644 pp.

J. A. Steers, V. J. Chapman, J. Colman and J. A. Lofthouse (1940), 'Sand cays and mangroves in Jamaica', *Geographical Journal*, 96: 305–28.

A. Strahan (1902), 'On the origins of the river system of South Wales and its connection with that of the Severn and Thames', *Quarterly Journal of the Geological Society of London*, 58: 207–25.

W. D. Thornbury (1954), *Principles of Geomorphology*. 618 pp.

F. M. Trotter (1929a), 'The Tertiary uplift and resultant drainage of the Alston Block and adjacent areas', *Proceedings of the Yorkshire Geological Society*, 21: 161–80.

(1929b), 'The glaciation of eastern Edenside, the Alston Block and the Carlisle Plain', *Quarterly Journal of the Geological Society of London*, 85: 549–612.

A. E. Trueman (1938), *The Scenery of England and Wales*. 351 pp. (2nd edn as *Geology and Scenery in England and Wales* 1949, 349 pp.)

H. C. Versey (1937), 'The Tertiary history of East Yorkshire', *Proceedings of the Yorkshire Geological Society*, 23: 302–14.

L. R. Wager (1937), 'The Arun River drainage pattern and the rise of the Himalayas', *Geographical Journal*, 89: 239–49.

A. Wegener (1915), *Die Entstehung der Kontinente und Ozeane*. (2nd edn 1920, 3rd edn 1922, 4th edn 1929; 3rd edn translated as *The Origin of Continents and Oceans*, 1924.)

L. J. Wills (1924), 'The development of the Severn Valley in the neighbourhood of Iron-bridge and Bridgnorth', *Quarterly Journal of the Geological Society of London*, 80: 274–314.

(1929), *The Physiographical Evolution of Britain*. viii, 376 pp.

(1938), 'The Pleistocene development of the Severn from Bridgnorth to the sea', *Quarterly Journal of the Geological Society of London*, 94: 161–242.

S. W. Wooldridge (1927), 'The Pliocene history of the London Basin', *Proceedings of the Geologists' Association*, 38: 49–132.

(1928), 'The 200-foot platform in the London Basin', *Proceedings of the Geologists' Association*, 39: 1–26.

(1938), 'The glaciation of the London Basin and the evolution of the Lower Thames drainage system', *Quarterly Journal of the Geological Society of London*, 94: 627–67.

(1949), 'On taking the ge- out of geography', *Geography*, 34: 9–18.

(1951), 'Progress in geomorphology', in G. Taylor (ed.), *Geography in the Twentieth Century: a study of growth, fields, techniques, aims and trends*, 165–77.

(1952), 'Reflections on regional geography in teaching and research', *Transactions and Papers of the Institute of British Geographers*, 16: 1–11.

(1955), 'The study of geomorphology', *Geographical Journal*, 121: 89–90.

(1958), 'The trend of geomorphology', *Transactions of the Institute of British Geographers*, 25: 29–35.

S. W. Wooldridge and J. F. Kirkaldy (1936), 'River profiles and denudation-chronology in southern England', *Geological Magazine*, 73: 1–16.

S. W. Wooldridge and D. L. Linton (1938), 'Influence of Pliocene transgression on the geomorphology of south-east England', *Journal of Geomorphology*, 1: 40–54.

(1939), *Structure, Surface and Drainage in South-east England (Transactions of the Institute of British Geogaphers)*, 10: i–xvi, 1–124. (2nd edn, 1955, 176 pp.)

S. W. Wooldridge and R. S. Morgan (1937), *The Physical Basis of Geography: an outline of geomorphology*. xxi, 445 pp. (2nd edn as *Geomorphology: the physical basis of geography*, 1959, 409 pp.)

W. B. Wright (1914), *The Quaternary Ice Age*. 464 pp. (2nd edn. 1937, 478 pp.)

F. E. Zeuner (1938), 'The chronology of Pleistocene sea-levels', *Annals and Magazine of Natural History*, (11) 1: 389–405.

12 British geography, 1918–1945: a personal perspective

J. A. PATMORE*

To the majority of contemporary geographers, the contribution of the years covered by this book will seem little more than, at best, an historical footnote to the infinitely more voluminous and relevant material of the succeeding four decades. For the undergraduate in particular, the concepts and the names which have surfaced will have scant significance save as grist for the mill in the historical sections of the near ubiquitous 'principles' or 'general' paper. Even then, it is depressingly rare for the work quoted to have been read in original form rather than in abstract in a later commentary. Indeed, one of the most respected of those commentaries (Johnston 1983) itself takes 1945 as its initial point of reference.

Those years, however, have far more than antiquarian value, as the most cursory reading of the chapters of this volume bears witness. They laid the effective foundations of university teaching in the discipline, spanning its emergence as an honours degree subject in its own right to its acceptance as a core subject in any credible university curriculum. They nourished a conceptual and intellectual framework which still has relevance despite the ferment and the fruits of more recent years. They nurtured a fellowship in which personal relationships had a significance beyond academic intercourse, and gave geography a fervour and freshness which underpinned its intellectual attractions, and which happily in large measure it still retains.

In none of these contexts do the contributions in this book need any garnish. They are individual, even idiosyncratic, but in sum they bring a fresh and worthy recognition of the inter-war legacy. This concluding

---

* John Allan Patmore (b. 14 November 1931) graduated in the Honour School of Geography in the University of Oxford in 1952. He taught in the department of geography in the University of Liverpool from 1954–73. In 1973 he was elected to the Chair of Geography in the University of Hull.

comment is in no sense an independent, *post hoc* assessment of that legacy, but a personal reflection on the lineaments the chapters have revealed.

It is inevitably pointed by personal perception. At the end of the period, four decades ago, the writer was in the middle of his secondary school course. Indeed, 1945 is best remembered as the year in which geography was dropped from his personal curriculum in favour of history, and only fully restored on his decision in 1949 to read geography at university, despite an entrance scholarship in history and a serious college warning as to the probable consequences! But if his training was entirely post-1945, it was rooted in the concepts and contributions of the inter-war years. The depth of those roots has been emphasized again in reading these essays. Names emerge to happy recollection; Eva Taylor, with waspish delight demolishing the population predictions of a hapless young Bracknell planner; and S. W. Wooldridge, eyebrows bristling, supporting with personal vigour and academic rigour the research conclusions of one of his postgraduates against the questioning of a non-King's geomorphologist.

Even more, the roll of contributors is a roll of personal friends and mentors. For three of these, the contribution is unhappily their last published work, for they died shortly after its completion: Emrys Bowen, with nonconformist fervour, always happy to show that every worthwhile innovation had its roots in Aberystwyth; Kenneth Edwards, a patient and courteous external examiner of a postgraduate thesis, always anxious to maintain the link with the student and kindly monitor progress over succeeding years; Stanley Beaver, relishing a shared enthusiasm for railways, yet beyond the enthusiasm encouraging an academic perspective and concern. With such memories, academic detachment becomes difficult if not impossible. Nevertheless, a number of themes emerge from the collection, themes which encapsulate something at least of the continuing legacy of the period.

The first, pervasive, theme is the extremely modest nature of the resources deployed, in both human and financial terms. In this respect, of course, geography was not unique among university disciplines, but it is salutary to recall the size of typical teaching staffs. Even the bigger London departments were of modest scale. Beaver recalls that at L.S.E. in the 1930s there were five members of staff, though L.S.E. was only responsible for the economic and regional aspects of the Joint School degree course. In the University Colleges teaching the London syllabus, such provision would have seemed generous indeed, and typically only one or two members of staff handled the whole of the work. At Southampton in the 1920s, for example, O. H. T. Rishbeth and Miss F. C. Miller carried the entire burden of teaching. The pressure was the greater in that the syllabus was externally determined, without that freedom of topic which the contemporary university teacher takes so much for granted.

Despite Herculean endeavours, it was often difficult to offer a fully balanced, integrated course. Individual members of staff showed a flexibility few would care to emulate today. Even where resources permitted specialization, there was rarely rigid demarcation. Some teachers were polymaths of a high order. L. Dudley Stamp, for example, held chairs successively in geology, economic geography and social geography. Others were concerned to use the wider implications of their subject. S. W. Wooldridge, with an incomparable reputation in geomorphology and the joint author of a highly respected standard text in physical geography (Wooldridge and Morgan 1937), was also concerned with the wider impact of the effect of the physical environment on the development of human settlement, as his contribution to Darby's seminal text on historical geography (Darby 1936) exemplifies.

Even this flexibility of mind and breadth of vision could not wholly compensate for the simple lack of human resources. Even in the 1930s, school examination syllabuses espoused a reasonable balance between human and physical geography as well as the usual substantial injection of regional knowledge. That same balance was not always as evident at university level. Human and regional aspects were usually well covered, often, as at Aberystwyth, with a very distinctive flavour. For physical geography, as Steers recalls, the picture was very different. Several geography departments had emerged from joint departments of geography and geology, and physical geologists continued to have a very important role in the teaching of geomorphology. That influence lasted well into the recent period: the writer remembers with great pleasure the distinctive contribution of K. S. Sandford at Oxford in the 1949–50 session before the arrival of M. M. Sweeting to stiffen teaching in that area. Apart from geomorphology, most courses contained at least the rudiments of meteorology and climatology, but there was often less effective cover of biogeography apart from simple descriptions of vegetation distribution.

The second theme links closely to the first, but at first glimpse the link is not an obvious corollary. With energies so fully absorbed in teaching, the time available for research and writing would seem inevitably limited yet the period is marked by a sustained quality of scholarly output. Perhaps in these very limitations is the engine for activity. Darby certainly perceives it to be so:

We had something of the dogmatic fervour of new converts to a faith, heightened by the fact that the position of geography as an academic discipline was not all that well-established. Being insecure, we were emphatic.

We were, moreover, dissatisfied with professing the new faith without attempting good works (see above, p. 124).

That fervour was not confined to historical geography. Research perhaps had its surest ground where the techniques had already been honed in allied disciplines. E. G. R. Taylor's work on the history of English geographical thought from 1485 to 1650 (Taylor 1930; 1934) broke new ground but it depended essentially on the scholarly use of the muniment room which had long been the historian's stock-in-trade.

Other research enterprises were remarkable as much for the energy as the vision they displayed. Fitting pride of place must go to Stamp's Land Utilisation Survey. By any standards, the work was monumental. Its basic lineaments bear simple repetition: the surveying of some 22,000 six-inch quarter sheets, the reduction of the work to one-inch scale, and the production of ninety-two county reports. As Willatts has recorded, the staff was miniscule for the scale of the enterprise, but was sustained not only by Stamp's unbounded enthusiasm and energy but by his practical organizational skills and fund-raising ability. Perhaps not even Stamp foresaw the wider impact that the Survey would have. It had importance not only as a record, but as a description of land use '... at a very distinctive period ... when ... agricultural use of land was at a nadir, when the proportion of the surface under the plough had reached the lowest ever recorded since statistics were first collected in 1866' (Stamp 1947: 404). Even more important, its timing made it a fundamental document for the post-1945 planning ferment, 'an obvious basis for replanning Britain' (Freeman 1980: 136).

In a more academic context, the needs of teaching in a burgeoning discipline brought a remarkable spate of writing, the production of texts which not only served their period well but which were the foundation of teaching in the post-war years. The pace of change in the real world meant that some had a relatively limited life. The land portrayed in *Great Britain: essays in regional geography* (Ogilvie, 1928) was a transient land, the book 'a testimony to the Britain and to the geographers of its period' (Mitchell 1962: xi). Nonetheless, the impact of the book was such that the editor of the successor volume had specifically to declare that 'this collection of essays is not a new edition' of its precursor. Other books had such a stature and a compass that successive revisions kept them in the forefront. Perhaps one of the most remarkable was Stamp and Beaver's *The British Isles: a geographic and economic survey*, first published in 1933, reaching a sixth edition in 1971 and still in print in its golden jubilee year. Many other works of the period retain more than an antiquarian interest – Darby on historical geography (Darby 1936), or Taylor on the history of geography (Taylor 1930; 1934) to name but two which figure prominently in this book. Even in fields more marked by change, or in fields more recently emergent, it remains surprising how often books and papers of the period are prescient in perception, laying foundations and providing the stimulus for the work of more recent years. The geography of recreation, for example, had little form before the late 1960s, but Gilbert's important paper on holiday resorts (Gilbert 1939) remains a valuable reference.

The third theme indeed would emphasize continuity. The geography of the 1918–45 period is noteworthy not only for what it did achieve with a fraction of the resources the subject can currently command, but for the impetus it has given to post-war change. To some extent that is expected and inevitable. Most of the authors of this book, for example, were trained in the pre-war world, but all made their major contributions in the post-war era. The links, however, are not always quite so obvious. In geomorphology, work on the explanatory description of landforms was pioneered between the wars: Wooldridge and Linton's monograph on *Structure, Surface and Drainage in South-east England* (1939) pointed the way to effective interpretation, and more than justified its second edition in 1955. Unfortunately, the breadth of grasp of such progenitors did not always infuse later work. In the words of a recent commentator, 'we became fascinated by what seemed to be evidence of an episodic development of landform, and turned from the reality of major upland surfaces to the seduction of fragmented flats surveyed at ever closer intervals, yet interpreted in terms of the same ideas. ... It is all rather sad, for the grand outline was properly understood' (Clayton 1980: 174).

The reaction against the narrow concerns of denudation chronology came in the search for an understanding of geomorphological processes on a new and generally much smaller scale. That search had already had its antecedents, and Steers traces the role of researchers at Cambridge in coastal and glacial work. While they antedated the sophisticated application of quantitative techniques, they depended upon accurate measurement in the field and fully recognized the need for precise, surveyed data rather than intuitive reconnaissance.

The impetus for change came in the applied as well as the pure aspects of the subject. The upsurge of planning in post-war Britain, fuelled in particular by the 1947 Town and Country Planning Act, brought the need for a new generation of planners with wider spatial skills than those possessed by architects alone. Development plans were concerned with techniques of survey and of integrated appraisal which were the familiar stuff of geography. The entry of geographers in growing numbers into the planning profession was stimulated by the recognition of the worth of the existing geographical contribution. Initially that recognition was neither sought nor fostered: in the Land Utilisation Survey, as Willatts recalls, 'we were too involved in the completion of our publication programme to be able to spare much effort to bring our work to the attention of organizations outside the academic field'. But that attention soon came, simply because the worth of geographical survey was seen in the results it achieved. The Land Utilisation Survey itself, the work of G. H. J. Daysh and A. A. L. Caesar in the North-east, and the contributions of geographers to the Barlow, Scott and Uthwatt Reports sparked a growing involvement, and the establishment of the infant Ministry of Town and Country Planning in 1943 saw geographers fully and actively employed, and set the pattern for the new conditions of the post-war era.

The fourth theme needs brief treatment only, but in a sense it arises almost despite these essays. The overall theme of the collection was the 'issues and ideas' of the period, but time after time, overtly and covertly, the issues are suborned by the personalities. The academic geographers of the inter-war years were a tight-knit fellowship, little more than a hundred in number even in 1945. For many, working in departments with only one or two staff, stimulus came not only in academic converse with colleagues in other disciplines, but especially in the meetings of the major geographical bodies – the Royal Geographical Society, the Geographical Association, and the particular creature of the period and of the university geographer, the Institute of British Geographers (Steel 1984) of 1933.

Many contributed a great deal of their time to the administration and the academic fellowship of these bodies. Bowen has chronicled Fleure's succession to both the Honorary Secretaryship and the Honorary Editorship of the Geographical Association. Edwards recalls with evident affection that 'like most academic geographers I gave regular support to the Geographical Association. It was our own organisation ... I responded to its needs whenever I could'. For the Association at least, the fellowship was not that of the university teacher alone, but of the secondary school teacher as well. The integration of teachers at all levels helped to create a unity of spirit and of purpose which infused geography more than most academic disciplines. It is all the more to be regretted that growing numbers, intensifying specialization, and the increasing separation of teachers in differing institutions have not only changed the pattern, if not the intensity, of geographical fellowship, but have weakened the sense of single, simple identity.

The final theme is far more a personal testament and legacy. As has already been made clear, the writer was trained outside the period which is the concern of this book, but in a tradition and by teachers wholly rooted in it. These essays therefore have prompted the insistent question as to the nature of the legacy, if any, which the period has given to a contemporary practitioner. The answer is obviously personal, and may reflect personal perceptions and prejudices as much as any conscious legacy. Some facets indeed are far from unique to the period in question. This is certainly true for two of the most persistent themes, the sense of corporate fellowship and the sense of structural unity, or rather structural wholeness, in the discipline. The fellowship and the unity may have had roots in smallness of scale rather than inherent values but they are prized with affection if not always practised with vigour.

A further thread perhaps betrays its roots more clearly, the sense of place. The geography of the period was nothing if not regionally oriented. Bowen notes the heavy regional bias of the Aberystwyth degree, but such a balance was far from unique to the Aberystwyth course. At Oxford in the early 1950s, half of the Final Honour School papers were specifically regional, and to those must be added the individual regional description. But the term 'regional' was something of a misnomer. Regional concepts as such had little emphasis at the undergraduate level, whether the elegant global constructs of Herbertson or the more arid local formulations of Unstead. But a sense of place most certainly had. Regional geography was concerned with a very real world. Its description might at times border on the tedious, but the sense of reality was pervasive and enlivening. Geography was concerned with the tangible landscape, and an eye for country was the most valued of the geographer's tools.

Perception alone, of course, was not enough: understanding, however partial, was a concurrent aim. Yet the search for understanding came to prize the mechanism beyond the place: too much of the geography of the last forty years has been concerned to create theoretical frameworks too far removed from the realities of the world. For this geographer at least, a sense of place inculcated by practice rather than by precept has been the happiest legacy of his formative years, a gift which continues to infuse his geography with purpose and with pleasure. Happily, the signs are that others too prize that legacy. R. J. Johnston has called for geographers 'to remove their blinkers, to cast off their parochial myopia and once again teach about the world as a mosaic of places, not as a series of examples of von Thünen and Christaller' (Johnston 1984: 445).

To heed such a call builds geography firmly on the roots that this volume, however hesitantly, delineates.

REFERENCES

K. M. Clayton (1980), 'Geomorphology' in E. H. Brown (ed.), *Geography Yesterday and Tomorrow*, 167–78.
H. C. Darby (1936) (ed.), *An Historical Geography of England before AD 1800*.
T. W. Freeman (1980), *A History of Modern British Geography*.
E. W. Gilbert (1939), 'The growth of inland and seaside health resorts in England', *Scottish Geographical Magazine*, 55, 16–35.
R. J. Johnston (1983), *Geography and Geographers: Anglo-American human geography since 1945*.
   (1984), 'The world is our oyster', *Transactions of the Institute of British Geographers*, 9.4: 441–59.
J. B. Mitchell (1962) (ed.), *Great Britain: geographical essays*.
A. G. Ogilvie (1928) (ed.), *Great Britain: essays in regional geography*.
L. D. Stamp (1947), *The Land of Britain: its use and misuse*.
R. W. Steel (1984), *The Institute of British Geographers: the first fifty years*.
E. G. R. Taylor (1930), *Tudor geography, 1485–1583*.
   (1934), *Late Tudor and Early Stuart geography, 1583–1650*.
S. W. Wooldridge and R. S. Morgan (1937), *The Physical Basis of Geography: an outline of geomorphology*.

Index

(The index has been prepared by Eileen M. Steel)